SPACE EXPLORATION
A Journey Through the Cosmos

SPACE EXPLORATION
A Journey Through the Cosmos

Austin Mardon, PhD

MERCURY LEARNING AND INFORMATION
Boston, Massachusetts

MERCURY LEARNING AND INFORMATION
121 High Street, 3rd Floor
Boston, MA 02110
info@merclearning.com

A. Mardon. *Space Exploration: A Journey Through the Cosmos*
ISBN: 978-1-50152-368-7

Library of Congress Control Number: 2024949169

242526321   This book is printed on acid-free paper in the United States of America.

*To Bill Cassidy, my Antarctic meteorite expedition leader.
A friend who with me and who searched for shattered
stars on the frozen seas beyond the South
Pole when I was a young man.*

CONTENTS

PREFACE

INSIGHTS INTO THE SIGNIFICANCE OF SPACE EXPLORATION

Space exploration is one of humanity's most remarkable endeavors, reflecting our deepest desires to understand the universe and our place within it. Its significance extends far beyond technical achievements and scientific discoveries; it profoundly impacts our society, culture, and future.

At its core, space exploration embodies human curiosity and the relentless pursuit of knowledge. From early stargazing to today's sophisticated missions, our drive to explore the cosmos has pushed the boundaries of what we know and can achieve. This curiosity has fueled countless innovations, leading to technological advancements that transform our daily lives. For example, satellite technology has revolutionized global communication, weather forecasting, and navigation, making the world more connected and informed.

The dawn of the space age marked a pivotal moment in human history. The Soviet Union's launch of Sputnik 1 in 1957 ignited wonder and competition, propelling humanity into a new era. The subsequent space race between the United States and the Soviet Union was not just a display of technological prowess but a symbol of human potential and determination. Achievements such as Yuri Gagarin's historic flight and the Apollo 11 moon landing captured the world's imagination and showed us that the seemingly impossible could be made possible.

Space exploration has always been a powerful catalyst for international cooperation. The construction and operation of the International

Space Station (ISS) exemplify what can be achieved when nations collaborate toward a common goal. The ISS serves as a platform for scientific research, technological development, and educational outreach, fostering a spirit of unity and shared purpose. It reminds us that the challenges of space are best met through cooperation rather than competition, and that the benefits of exploration extend to all humanity.

Beyond tangible advancements, space exploration profoundly impacts our understanding of ourselves and our place in the universe. Images of Earth from space, like the iconic "Blue Marble" photo taken by the Apollo 17 crew, shift our perspective, highlighting our planet's fragility and interconnectedness. These views underscore the need for environmental stewardship and inspire a sense of global responsibility.

The pursuit of space exploration also drives us to confront fundamental questions about existence and the potential for life beyond Earth. Missions to Mars, the moons of Jupiter and Saturn, and distant exoplanets are not just scientific endeavors; they seek to answer the age-old question of whether we are alone in the universe. The search for extraterrestrial life challenges our understanding of biology, chemistry, and the conditions necessary for life, broadening our horizons and expanding the scope of human inquiry.

As we look to the future, the significance of space exploration becomes even more apparent. The challenges we face on Earth, such as resource scarcity, environmental degradation, and geopolitical tensions, underscore the need for sustainable solutions and new frontiers. Space offers vast resources and opportunities for technological innovation to address these challenges. Concepts like asteroid mining, space-based solar power, and the colonization of other planets hold the promise of ensuring humanity's long-term survival and prosperity.

Moreover, the quest for interstellar travel embodies the ultimate expression of human aspiration. Projects like the Breakthrough Starshot initiative aim to send probes to the nearest star systems, pushing the boundaries of our current technological capabilities. These ambitious endeavors inspire a new generation of scientists, engineers, and dreamers to push the limits of what is possible and envision a future where humanity explores and inhabits the stars.

In conclusion, space exploration is far more than a series of missions and discoveries. It reflects our deepest aspirations, capacity for innovation, and desire to understand the universe. It challenges us to think beyond our immediate surroundings and consider the broader implications of our actions. As readers embark on "A Journey Through Space Exploration," I hope they gain a deeper appreciation for the significance of this incredible journey and feel inspired to contribute to humanity's ongoing quest to explore the cosmos.

CHAPTER 1

THE HUMAN FASCINATION WITH THE COSMOS

INTRODUCTION

From time immemorial, humans have gazed at the stars with wonder and curiosity. The vast expanse of the night sky, dotted with luminous celestial bodies, has inspired myths, legends, and scientific inquiries. This chapter delves into the early astronomical discoveries and ancient space myths that have shaped our understanding of the cosmos. By exploring the ways in which different civilizations have interacted with the night sky, readers gain insight into the universal human drive to comprehend their place in the universe. Ancient peoples used the heavens as a vast canvas, creating stories and myths that explained natural phenomena and articulated their cosmologies. From the meticulous astronomical records of the Babylonians to the mythic narratives of the Greeks and the precise calendrical systems of the Maya, this chapter examines how early astronomical practices not only reflected on but also shaped the worldviews and daily lives of ancient cultures.

EARLY ASTRONOMICAL DISCOVERIES

1. **Question:** What is cosmology, and how did ancient civilizations engage with it?

   Answer: Cosmology is the study of the universe's origins, structure, and nature. Ancient civilizations engaged with cosmology through myths, legends, and observations of celestial bodies to explain natural phenomena and their place in the cosmos. These civilizations used the sky as a canvas, weaving stories that connected the stars to their deities, heroes, and historical events. For instance, the ancient Greeks personified celestial bodies, giving them characteristics and narratives that explained their movements and significance. The Chinese developed intricate lunar calendars and documented celestial events meticulously, seeing them as omens or divine messages.

2. **Question**: Which civilization is credited with developing one of the earliest known calendars?

   Answer: The Mayan civilization is credited with developing one of the earliest known calendars, called the Haab' and Tzolk'in calendars. The Maya combined these two calendars to create the Long Count calendar, which tracked time over long periods. The Haab' was a 365-day solar calendar, while the Tzolk'in was a 260-day ritual calendar. Their ability to predict solar and lunar eclipses and other celestial events demonstrated their sophisticated understanding of astronomy. The precision and complexity of the Maya calendar system reflect their advanced mathematical skills and deep astronomical knowledge. (See Figure 1.1.)

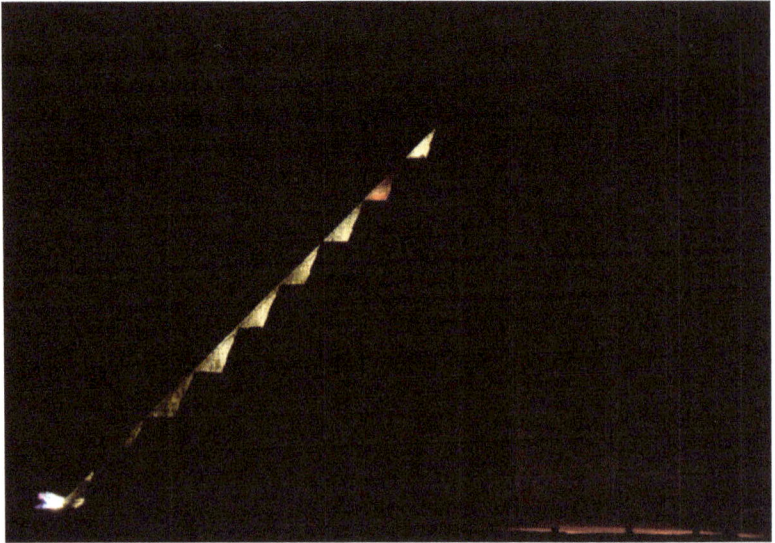

FIGURE 1.1 The pyramid at Chichén Itzá during the equinoxes, where the shadow creates the illusion of a serpent slithering down the steps. This demonstrates the ancient Maya's sophisticated understanding of solar movements. Source: Wikimedia Commons.

3. **Question**: What role did astronomy play in ancient Egypt?
    Answer: Astronomy played a crucial role in ancient Egypt, influencing the design of temples and pyramids aligned with celestial events like solstices and equinoxes. The annual flooding of the Nile River, vital for agriculture, was predicted using the heliacal rising of Sirius. This event marked the beginning of the Egyptian New Year and was celebrated with festivals. The alignment of the pyramids and other structures with specific stars and constellations reflected the Egyptians' belief in the divine nature of the cosmos and its influence on earthly affairs. Their astronomical knowledge was integral to their religious and agricultural practices.

4. **Question**: How did ancient Greeks contribute to our understanding of astronomy?
    Answer: Ancient Greeks made significant contributions to astronomy, including the development of the geocentric model by Ptolemy, Aristotle's studies on celestial bodies, and Aristarchus's suggestion of a heliocentric system. Ptolemy's Almagest compiled the knowledge of Greek astronomy, presenting a complex model of the universe with Earth at its center. Aristotle's work laid the foundation for understanding the natural world, including the motion of celestial bodies. Aristarchus proposed that the sun, not Earth, was at the center of the universe, a revolutionary idea that predated Copernicus by almost two millennia. Greek philosophers and astronomers used mathematical principles to explain celestial phenomena, influencing future generations of scientists.

5. **Question**: What was the Babylonian contribution to astronomy?
    Answer: The Babylonians developed a sophisticated system of recording planetary motions, enabling them to predict lunar and solar eclipses accurately. They used a base-60 (sexagesimal) number system, which facilitated their complex astronomical calculations. The Babylonians compiled extensive records on clay tablets, documenting the positions and movements of celestial bodies over long periods. These records, known as the Astronomical Diaries, were used to create predictive models for eclipses and other celestial events. The Babylonians' detailed observations and mathematical methods laid the groundwork for later astronomical advances in the Hellenistic world and beyond.

6. **Question**: What celestial object did the ancient Chinese recognize and record first?
   Answer: The ancient Chinese recognized and recorded comets first, noting their appearances and characteristics as early as 2000 BCE. They viewed comets as omens, often associating them with significant earthly events such as the rise and fall of rulers or natural disasters. Chinese astronomers meticulously documented the positions, movements, and physical appearances of comets, which allowed them to develop early theories about their nature. These records were part of a broader tradition of celestial observation that included the tracking of planetary movements, eclipses, and the development of detailed star catalogs.

7. **Question**: How did ancient Indian texts describe the cosmos?
   Answer: Ancient Indian texts like the Rigveda and Surya Siddhanta describe the cosmos through metaphors, mathematical calculations, and early models of planetary motion, hinting at advanced understanding. The Rigveda contains hymns that speak of the creation of the universe and the movements of celestial bodies. The Surya Siddhanta, an astronomical treatise, presents mathematical methods for calculating the positions of planets, the duration of eclipses, and other astronomical phenomena. These texts reflect a deep integration of astronomy with religious and philosophical thought, emphasizing the interconnectedness of the cosmos and human life.

8. **Question**: Which ancient civilization created the zodiac signs?
   Answer: The Babylonians created the concept of the zodiac signs based on twelve constellations along the ecliptic. These constellations corresponded to the path that the sun, moon, and planets appear to follow over the course of a year. The Babylonians used the zodiac to predict celestial events and interpret their significance for human affairs. This system was later adopted and refined by the Greeks, who added their own mythological interpretations to the constellations. The zodiac remains a foundational element of astrology and continues to influence both astronomical and cultural traditions.

9. **Question**: How did ancient Indian astronomers contribute to our understanding of the cosmos?
   Answer: Ancient Indian astronomers, such as Aryabhata and Brahmagupta, made substantial contributions by calculating the Earth's circumference, theorizing heliocentric models, and accurately predicting solar and lunar eclipses. Their works influenced both Eastern and Western astronomical traditions.

10. **Question**: What is Stonehenge, and what is its significance in astronomy?

    Answer: Stonehenge is a prehistoric monument in England. Its stones are aligned with solar phenomena, like the summer solstice sunrise, suggesting its use as an ancient astronomical observatory. The precise alignment of the stones with the solstices indicates that the builders had a sophisticated understanding of the solar cycle. Stonehenge may have served as a calendar, helping to mark the changing seasons and important agricultural dates. Additionally, it likely held ceremonial and ritual significance, possibly serving as a site for gatherings and celebrations linked to celestial events.

11. **Question**: What significant ancient astronomical contributions did China make?

    Answer: Ancient China made significant contributions to astronomy, including the earliest known star catalog by Shi Shen, the recording of supernovae, and the creation of detailed celestial maps. These efforts laid the groundwork for future astronomical studies.

12. **Question**: What was the significance of the Dunhuang Star Atlas in Chinese astronomy?

    Answer: The Dunhuang Star Atlas is one of the oldest known complete star maps, dating back to the Tang Dynasty (618–907 AD). It documents over 1,300 stars and shows the precision with which ancient Chinese astronomers mapped the night sky. (See Figure 1.2.)

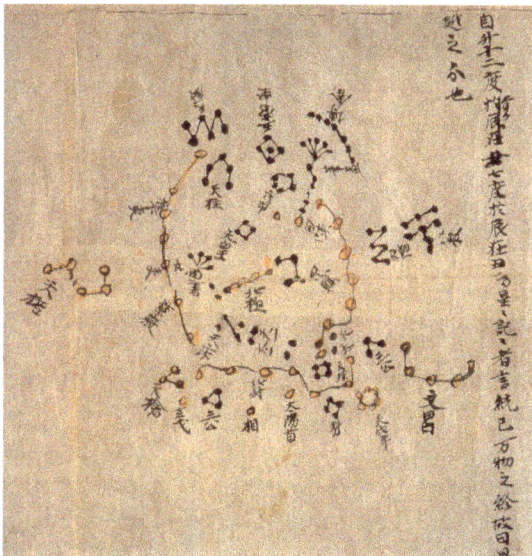

FIGURE 1.2 The Dunhuang Star Atlas, the oldest known complete star atlas, dating from the seventh century, maps over 1,300 stars and includes notable constellations like the Big Dipper. Source: Wikimedia Commons.

13. **Question**: How did the Aztecs view the cosmos?
    Answer: The Aztecs viewed the cosmos as cyclical, with their calendar systems reflecting cosmic order. They believed that celestial movements influenced daily life and required rituals for cosmic balance. The Aztec calendar consisted of a 365-day solar calendar (the Xiuhpohualli) and a 260-day ritual calendar (the Tonalpohualli). The intersection of these cycles created a fifty-two-year period known as the "century." The Aztecs performed various rituals and ceremonies to ensure the continued harmony of the universe, including sacrifices to the gods who controlled celestial phenomena. Their cosmology was deeply intertwined with their religious and social practices.

14. **Question**: How did the ancient Indian text Surya Siddhanta influence astronomical understanding?
    Answer: The Surya Siddhanta, an ancient Indian astronomical text, provided detailed descriptions of planetary motions, eclipses, and timekeeping methods. It significantly influenced Indian astronomy and was later translated into Arabic and Greek, impacting global astronomical knowledge.

ANCIENT SPACE MYTHS

15. **Question**: Which ancient philosopher is known for the idea of the celestial sphere?
    Answer: Eudoxus of Cnidus, an ancient Greek philosopher, is known for conceptualizing the celestial sphere. He proposed that the heavens were composed of concentric spheres with Earth at the center. Each sphere carried a celestial body, and their rotations explained the observed movements of the stars and planets. This model was influential in the development of Greek astronomy and laid the groundwork for the geocentric models of Aristotle and Ptolemy. The idea of the celestial sphere persisted in Western astronomy until the Renaissance when it was replaced by heliocentric models.

16. **Question**: What role did constellations play in ancient cultures?
    Answer: Constellations served various purposes in ancient cultures, such as guiding navigation, marking agricultural seasons, and inspiring myths about gods and heroes. For example, the ancient Greeks used constellations to navigate the seas and created

mythological stories around them, such as the tale of Orion the Hunter. The Egyptians aligned their pyramids and temples with specific stars to reflect their religious beliefs. In many cultures, constellations were seen as celestial markers that indicated the timing of planting and harvesting crops. These star patterns also provided a framework for storytelling and cultural identity.

17. **Question**: What are the Vedic texts, and how do they relate to astronomy?

    Answer: The Vedic texts are ancient Indian scriptures that contain hymns and descriptions of celestial phenomena, offering insights into early Hindu astronomy. The Rigveda, one of the oldest Vedic texts, includes hymns that reference the sun, moon, and other celestial bodies. The Surya Siddhanta, a later text, provides detailed mathematical formulas for calculating planetary positions, eclipses, and other astronomical events. These texts demonstrate the integration of astronomical knowledge with religious and philosophical teachings in ancient India, highlighting the importance of celestial observation in Vedic culture.

18. **Question**: How did Polynesians use astronomy in navigation?

    Answer: Polynesians used celestial navigation, relying on star positions, moon phases, and the rising and setting of specific stars to guide them across the Pacific. They developed a sophisticated understanding of the night sky, using star charts known as "star compasses" to navigate vast ocean distances. Polynesian navigators, or wayfinders, memorized the positions of hundreds of stars and their movements throughout the year. This knowledge enabled them to undertake long voyages between islands, using the stars as reliable guides even when out of sight of land.

19. **Question**: Which Mesopotamian epic reflects ancient views of the cosmos?

    Answer: The Epic of Gilgamesh reflects Mesopotamian views of the cosmos, with its references to celestial beings, the sun god Shamash, and cosmic cycles. This ancient epic, one of the earliest known works of literature, portrays the cosmos as a realm governed by powerful deities who control natural and celestial phenomena. The journey of Gilgamesh, the hero, is intertwined with encounters and guidance from these divine entities. The epic also illustrates the Mesopotamian belief in a structured universe, where celestial movements influenced earthly events and human fate.

20. **Question**: How did the Romans incorporate Greek astronomy into their culture?

    Answer: The Romans adopted Greek astronomical concepts, including Ptolemy's geocentric model, and integrated them into their calendars and astrology practices. Roman astronomers and astrologers built upon the work of Greek predecessors, translating and expanding upon their texts. The Roman calendar, reformed by Julius Caesar into the Julian calendar, was influenced by Greek astronomical knowledge. Astrology, deeply rooted in Greek traditions, became popular in Rome, where horoscopes and celestial interpretations played a significant role in daily life and decision-making.

21. **Question**: What was Hipparchus's major contribution to astronomy?

    Answer: Hipparchus, a Greek astronomer, compiled a star catalog and discovered the precession of the equinoxes. His star catalog, which listed the positions of over 850 stars, was one of the first systematic attempts to map the night sky. Hipparchus also identified the slow shift in the position of the equinoxes, a phenomenon now known as axial precession. This discovery was a significant advancement in understanding the long-term movements of celestial bodies and the mechanics of Earth's rotation. Hipparchus's work laid the foundation for future astronomical studies and was later incorporated into Ptolemy's Almagest.

22. **Question**: How did Islamic scholars preserve and expand upon ancient astronomical knowledge?

    Answer: Islamic scholars translated Greek and Indian texts, developed accurate instruments, refined star catalogs, and advanced mathematical models of planetary motion. During the Islamic Golden Age, scholars such as Al-Battani, Al-Sufi, and Ibn al-Haytham made significant contributions to astronomy. They built observatories, improved existing astronomical instruments like the astrolabe, and produced detailed star charts. Their work preserved and enhanced the knowledge of earlier civilizations, ultimately influencing European astronomy during the Renaissance. The translation movement in the Islamic world ensured that the astronomical heritage of Greece and India was transmitted and expanded upon.

23. **Question**: What was the purpose of the ancient astronomical observatory Jantar Mantar in India?

    Answer: Jantar Mantar, built in the eighteenth century, served to observe and record celestial movements with its large instruments and alignments. Constructed by Maharaja Jai Singh II in several cities, including Jaipur and Delhi, these observatories feature massive stone and metal structures designed for precise astronomical measurements. Instruments like the Samrat Yantra (a giant sundial) and the Jai Prakash (a hemispherical sundial) allowed astronomers to track the movements of the sun, moon, and planets with remarkable accuracy. Jantar Mantar reflects the advanced state of Indian astronomy and its integration with architectural innovation.

24. **Question**: Who is considered the father of modern astronomy, and why?

    Answer: Nicolaus Copernicus is considered the father of modern astronomy due to his heliocentric model, which placed the sun at the center instead of Earth. In his seminal work, "De revolutionibus orbium coelestium," Copernicus challenged the long-held geocentric model, proposing that the planets, including Earth, orbit the sun. This revolutionary idea laid the foundation for the scientific revolution, influencing subsequent astronomers like Kepler, Galileo, and Newton. Copernicus's heliocentric model provided a more accurate explanation of planetary motions and paved the way for a new understanding of the universe.

25. **Question**: What ancient text is known for describing eclipses with remarkable accuracy?

    Answer: The "Almagest" by Ptolemy describes lunar and solar eclipses with remarkable accuracy for its time. Ptolemy's work, a comprehensive treatise on astronomy, compiled and expanded upon the knowledge of previous Greek astronomers. The Almagest includes detailed mathematical models for predicting the occurrence and characteristics of eclipses, based on the geocentric model. Ptolemy's descriptions and calculations remained influential for over a thousand years, serving as the authoritative reference for astronomers in the Islamic world and medieval Europe.

26. **Question**: How did ancient Greek and Roman astrology influence modern astrology?

    Answer: Greek and Roman astrology established the zodiac signs and planetary interpretations, forming the foundation of modern astrological practices. The Greeks developed the concept of horoscopic astrology, where an individual's destiny was determined by the positions of celestial bodies at the time of their birth. The Romans adopted and adapted these practices, integrating them into their own cultural and religious framework. Many elements of modern astrology, including the twelve zodiac signs, planetary influences, and astrological houses, can be traced back to these ancient traditions.

27. Question: What is the significance of the Nazca Lines in astronomy?
Answer: The Nazca Lines are thought to align with celestial phenomena and possibly represent constellations, reflecting ancient Peruvian astronomical knowledge. These massive geoglyphs, created by the Nazca culture in the Peruvian desert, depict animals, plants, and geometric shapes. Some researchers believe that certain lines and figures are aligned with the positions of the sun, moon, and stars, suggesting their use as astronomical markers or calendars. The exact purpose of the Nazca Lines remains a subject of debate, but their potential astronomical significance highlights the advanced observational skills of the Nazca people. (See Figure 1.3.)

FIGURE 1.3 Aerial view of the Nazca Lines in Peru, showcasing their large scale and intricate designs. Source: Wikimedia Commons.

28. **Question**: What is the cosmological significance of the Popol Vuh?
    Answer: The Popol Vuh, a Mayan creation myth, offers a cosmological framework that describes the creation of the world and celestial movements. This sacred text of the K'iche' Maya recounts the creation of the universe, the gods' interactions, and the origins of humanity. It emphasizes the cyclical nature of time and the interconnectedness of the cosmos. The Popol Vuh reflects the Mayans' deep understanding of astronomical cycles, including the movements of the sun, moon, and Venus. The text provides valuable insights into the Mayan worldview and their interpretation of celestial events.

29. **Question**: How did the ancient Greeks measure the Earth's circumference?
    Answer: Eratosthenes measured the Earth's circumference using the angles of shadows cast in different cities at the same time. By comparing the shadow lengths in Alexandria and Syene (modern Aswan) during the summer solstice, Eratosthenes calculated the angle of the Earth's curvature. He then used the distance between the two cities to estimate the Earth's circumference with remarkable accuracy. His method, based on geometric principles, demonstrated the Greeks' advanced understanding of mathematics and astronomy. Eratosthenes's measurement was a significant achievement in the history of science, providing a foundation for future explorations of the Earth's dimensions.

30. **Question**: What celestial body was central to ancient Egyptian religious beliefs?
    Answer: The sun, personified as the god Ra, was central to ancient Egyptian religious beliefs and their daily rituals. The Egyptians revered the sun as the source of life and order, with Ra traveling across the sky by day and through the underworld by night. The journey of the sun god symbolized the cycle of life, death, and rebirth. Temples and monuments, such as the Great Pyramid of Giza, were aligned with the sun's movements, reflecting its importance in Egyptian cosmology. Solar deities like Ra, Horus, and Aten played key roles in Egyptian mythology and were central to their religious practices.

31. **Question**: What was the purpose of the Antikythera Mechanism?
    Answer: The Antikythera Mechanism, an ancient Greek device, calculated the positions of celestial bodies and predicted eclipses. Discovered in a shipwreck off the coast of the Greek island

Antikythera, this intricate mechanical device dates back to around 100 BCE. The mechanism's gears and dials allowed users to track the cycles of the sun, moon, and possibly the planets. It also included a calendar and could predict lunar and solar eclipses. The Antikythera Mechanism is considered one of the earliest known analog computers, showcasing the advanced technological and astronomical knowledge of ancient Greece.

32. **Question**: How did the Ancestral Puebloans use astronomy in Chaco Canyon?
   Answer: The Ancestral Puebloans built structures in Chaco Canyon aligned with solar and lunar cycles, possibly serving ritual and agricultural purposes. These ancient inhabitants of the American Southwest constructed buildings like the Sun Dagger, a rock formation that marks the solstices and equinoxes with light patterns. Other structures, such as the Great Kiva and the Pueblo Bonito, were aligned with cardinal directions and celestial events. The astronomical alignments suggest that Chaco Canyon was a center for both astronomical observation and ceremonial activities, reflecting the Puebloans' deep connection to the cosmos.

33. **Question**: What did the ancient Norse believe about the cosmos?
   Answer: The ancient Norse believed the cosmos was a giant tree called Yggdrasil, with different worlds connected through its branches. Yggdrasil, the World Tree, supported nine realms, including Asgard (home of the gods), Midgard (Earth), and Helheim (realm of the dead). The Norse cosmology included various celestial bodies, such as the sun and moon, personified as chariots driven by gods or giants. This mythological framework explained natural phenomena and the structure of the universe, emphasizing the interconnectedness of all things. The Norse sagas and Eddas preserve these cosmological beliefs, reflecting the culture's rich mythological tradition.

34. **Question**: How did the Greeks determine lunar cycles?
   Answer: Greek astronomers observed lunar cycles to create calendars and predict eclipses, refining these observations mathematically. They noted the phases of the moon and its regular 29.5-day cycle from new moon to full moon and back. By tracking the moon's position relative to the stars and its phases, Greek astronomers developed methods for predicting lunar and solar eclipses. Ptolemy's Almagest includes detailed descriptions of lunar cycles

and eclipse predictions, building on the work of earlier astronomers like Hipparchus. These observations were essential for timekeeping, navigation, and religious ceremonies.

35. **Question**: What does the term "heliacal rising" mean, and why
was it significant to ancient cultures?
Answer: The heliacal rising refers to the first visible appearance
of a star at dawn, just before sunrise. It was significant for marking
seasonal changes in agriculture and ritual events. Many ancient
cultures used the heliacal rising of specific stars to signal important
times of the year. For example, the ancient Egyptians observed the
heliacal rising of Sirius to predict the annual flooding of the Nile,
which was crucial for their agricultural cycle. The Maya also used
heliacal risings to align their calendar with agricultural and ceremonial events. This phenomenon provided a reliable astronomical
marker for timekeeping and planning.

36. **Question**: How did the ancient Greeks distinguish between planets and stars?
Answer: The Greeks called planets "wanderers" (from the Greek
word *planetes*) due to their unique movement compared to fixed
stars. While stars maintained relatively fixed positions in constellations, planets moved across the sky, sometimes appearing to reverse
direction in retrograde motion. Greek astronomers observed these
movements and noted that planets did not twinkle like stars. They
identified five visible planets: Mercury, Venus, Mars, Jupiter, and
Saturn. The study of these "wandering stars" led to the development of complex models to explain their motions, culminating in
Ptolemy's geocentric system.

37. **Question**: What astronomical structures did the ancient Mayans
build?
Answer: The ancient Mayans built observatories like El Caracol,
with strategic alignments to track solstices, equinoxes, and planetary movements. El Caracol, located in the city of Chichen Itza,
features windows aligned with the positions of Venus and the sun
during key times of the year. The Mayans also constructed pyramids and temples oriented to capture the light of the sun and
moon during solstices and equinoxes. These structures served both
astronomical and ceremonial purposes, reflecting the Mayans'
advanced understanding of celestial cycles and their importance in
religious rituals. (See Figure 1.4.)

FIGURE 1.4 El Caracol, also known as "The Snail," an ancient Mayan observatory with windows aligned to track the movements of Venus and other celestial bodies. Source: Wikimedia Commons.

38. **Question**: How did the Islamic scholar Al-Battani improve on Greek astronomy?

   Answer: Al-Battani refined Ptolemy's calculations, correcting the solar year length and contributing to the understanding of planetary orbits. His observations and mathematical adjustments resulted in more accurate predictions of celestial events. Al-Battani's work, preserved in his book *Kitab az-Zij*, included precise measurements of the solar year, lunar phases, and the positions of stars and planets. His contributions were later translated into Latin and influenced European astronomers during the Renaissance, demonstrating the continuity and enhancement of astronomical knowledge across cultures.

39. **Question**: How did ancient Egyptians calculate the length of the solar year?

   Answer: Ancient Egyptians calculated the solar year length using the heliacal rising of Sirius, observing its annual correlation with the Nile's flooding. By tracking the interval between successive heliacal risings of Sirius, they determined that the solar year was approximately 365.25 days long. This observation led to the development of the Egyptian calendar, which included a 365-day year with an additional day added every four years to account for the extra quarter day. The accuracy of their calendar was crucial for agricultural planning and religious festivals, underscoring the importance of astronomy in Egyptian society.

40. **Question**: What was the purpose of the ancient Chinese observatory, the Gaocheng Astronomical Observatory?

    Answer: The Gaocheng Astronomical Observatory was used to observe celestial movements and improve the accuracy of the Chinese calendar. Built during the Yuan Dynasty, this observatory featured large, precise instruments, such as the gnomon for measuring solar shadows and the armillary sphere for tracking star positions. The observations made at Gaocheng helped refine the Chinese lunisolar calendar, ensuring accurate predictions of eclipses and seasonal changes. The observatory's work reflected the Chinese emphasis on harmonizing human activities with celestial cycles and maintaining the emperor's mandate from heaven. (See Figure 1.5.)

FIGURE 1.5 This image shows some of the astronomical instruments from the Beijing Ancient Observatory, which has been used since the 1400s to observe celestial phenomena. Source: Wikimedia Commons.

41. **Question**: How did the Dogon people of West Africa describe the star Sirius?

    Answer: The Dogon people described Sirius and claimed it had an unseen companion star, known today as Sirius B, possibly based on ancient oral traditions. This remarkable knowledge, which aligns with modern astronomical findings, suggests a sophisticated understanding of the Sirius star system. The Dogon also associated Sirius with their creation myths, believing it to be the home of the Nommo, ancestral spirits who played a crucial role in their cosmology. The Dogon's detailed astronomical lore has fascinated researchers, sparking debates about the origins and accuracy of their celestial knowledge.

42. **Question**: What was the geocentric model of the universe?
    Answer: The geocentric model, supported by Ptolemy and Aristotle, posited that Earth was the center of the universe, with the sun and planets revolving around it. This model, detailed in Ptolemy's Almagest, included complex systems of epicycles and deferents to account for the observed retrograde motion of planets. The geocentric model dominated Western astronomy for over a millennium, shaping the medieval understanding of the cosmos. It was eventually challenged and replaced by the heliocentric model proposed by Copernicus, which placed the sun at the center of the solar system.

43. **Question**: How did the Sumerians represent constellations?
    Answer: The Sumerians represented constellations as symbols related to myths and seasonal events, forming the earliest known star charts. They identified groups of stars with deities and mythological figures, integrating these constellations into their religious and agricultural practices. Sumerian star charts, such as the "Three Stars Each" texts, organized the night sky into distinct regions, each associated with specific stars and constellations. These representations influenced subsequent Mesopotamian cultures, including the Babylonians, who further developed and refined the system of constellations.

44. **Question**: Which star was most crucial to early Islamic astronomy, and why?
    Answer: The star Polaris was crucial for navigation and determining the qibla (direction of Mecca) during prayers. Known as the North Star, Polaris's fixed position in the sky made it a reliable reference point for Islamic astronomers and navigators. Determining the qibla accurately was essential for Islamic religious practices, as Muslims needed to face Mecca during prayer. Islamic scholars developed sophisticated methods for calculating the qibla based on celestial observations, ensuring that this fundamental aspect of their faith was observed correctly.

45. **Question**: What ancient Polynesian constellation helped navigate the Pacific?

    Answer: The Polynesian constellation "Te Waka o Tamarereti," which includes the stars of Orion's Belt, helped navigate the Pacific. Polynesian navigators used the positions and movements of stars within this constellation to guide their voyages across vast ocean distances. The alignment of the stars provided a stable reference for direction, allowing navigators to maintain their course even when out of sight of land. This knowledge, passed down through generations, was crucial for the Polynesians' successful colonization of remote islands throughout the Pacific.

46. **Question**: What role did constellations play in ancient Mesopotamian agriculture?

    Answer: Mesopotamian farmers used constellations to time the planting and harvesting of crops, as they marked the changing seasons. Constellations such as the Pleiades were associated with specific agricultural activities, signaling the appropriate times for sowing and reaping. The Mesopotamians' careful observation of the night sky allowed them to develop a calendar system that aligned with the agricultural cycle, ensuring the success of their farming practices. This integration of astronomy with agriculture highlights the practical importance of celestial knowledge in ancient Mesopotamian society.

47. **Question**: What astronomical event was central to Aztec rituals?

    Answer: The appearance of Venus as the morning star was central to Aztec rituals, symbolizing the rebirth of the sun. The Aztecs closely observed Venus's movements and associated its heliacal rising with the god Quetzalcoatl. This event marked important ceremonies and offerings to ensure the favor of the gods and the continuation of cosmic order. The Venus cycle was integrated into the Aztec calendar system, guiding religious and agricultural activities. The significance of Venus in Aztec culture underscores the deep connection between celestial events and religious practices.

48. **Question**: What is the significance of the Mayan calendar's "Long Count"?

    Answer: The Mayan calendar's "Long Count" is significant for its detailed tracking of longer time periods, allowing the Mayans to predict celestial events over centuries. The Long Count calendar was used to record historical dates and project future events, spanning a cycle of approximately 5,125 years. It was based on a combination of smaller cycles, including the 260-day Tzolk'in and the 365-day Haab'. This system enabled the Mayans to synchronize their religious, agricultural, and social activities with celestial movements, reflecting their advanced understanding of time and astronomy.

49. **Question**: How did the ancient Greeks explain the motion of planets?
    Answer: The ancient Greeks explained the motion of planets through the geocentric model, with complex systems of epicycles and deferents to account for retrograde motion. According to this model, planets moved in small circles (epicycles) while simultaneously orbiting Earth on larger circles (deferents). This explanation, detailed in Ptolemy's Almagest, sought to reconcile the observed irregular movements of planets with the geocentric framework. Although intricate and eventually proven incorrect, the epicyclic model represented a significant attempt to understand and mathematically describe celestial mechanics.

50. **Question**: What role did the Pleiades star cluster play in ancient cultures?
    Answer: The Pleiades star cluster played a role in navigation, agricultural planning, and mythologies across various ancient cultures, including Greek, Maori, and Japanese traditions. In Greece, the Pleiades were associated with the planting and harvesting seasons, as their heliacal rising marked the start of the agricultural year. The Maori of New Zealand viewed the Pleiades, known as Matariki, as a symbol of renewal and the beginning of the new year. In Japan, the Pleiades, called Subaru, held cultural significance and were celebrated in festivals. These diverse cultural connections highlight the universal importance of the Pleiades in ancient astronomy and mythology.

51. **Question**: How did the Inca use astronomy in their architecture?
    Answer: The Inca used astronomy in their architecture by aligning structures like Machu Picchu with solstices and other significant celestial events. Buildings and ceremonial sites were oriented to capture the light of the sun during solstices and equinoxes, creating dramatic effects and marking important times of the year. For example, the Intihuatana stone at Machu Picchu was designed to align with the sun during the winter solstice, symbolizing the Inca's connection to the solar deity Inti. This integration of astronomical

knowledge with architectural design reflects the Inca's sophisticated understanding of the cosmos and its importance in their culture.

52. **Question**: What ancient text compiled the knowledge of Greek astronomy?
    Answer: The "Almagest" by Ptolemy compiled the knowledge of Greek astronomy, serving as a foundational text for centuries. This comprehensive work synthesized and expanded upon the contributions of earlier Greek astronomers, providing detailed models for predicting the positions and movements of celestial bodies. The Almagest included star catalogs, mathematical formulas, and descriptions of instruments, making it an essential reference for astronomers in the Islamic world and medieval Europe. Ptolemy's geocentric model, although eventually superseded, remained influential for over a thousand years.

53. **Question**: How did ancient African cultures use celestial bodies in their rituals?
    Answer: Ancient African cultures used celestial bodies to mark important rituals, seasonal changes, and navigation, with star patterns often linked to their mythology. For example, the Dogon people of Mali associated the star Sirius with their creation myths and agricultural cycles. The San people of southern Africa used the positions of stars to guide their movements and seasonal activities. Celestial observations were also integral to the rituals and ceremonies of the Yoruba and Ashanti peoples. These practices demonstrate the deep connection between astronomy and cultural traditions in ancient African societies.

54. **Question**: What is the significance of the Nebra Sky Disk?
    Answer: The Nebra Sky Disk, an ancient artifact from Germany, is significant for its depiction of astronomical phenomena, including the sun, moon, and stars, providing insights into Bronze Age astronomy. Discovered in 1999, the disk dates back to around 1600 BCE and features gold symbols representing celestial objects and possibly a solar boat. The disk is thought to have been used as a tool for observing and predicting celestial events, such as solstices. Its discovery has provided valuable information about the astronomical knowledge and practices of ancient European cultures, highlighting their sophisticated understanding of the cosmos.

The early astronomical discoveries and ancient space myths explored in this chapter reveal a profound connection between humans and the cosmos. These ancient observations and narratives laid the groundwork for modern astronomy, bridging the gap between myth and science. As has been seen, civilizations across the world—from the Babylonians and Greeks to the Maya and Chinese—developed sophisticated methods to track celestial phenomena and interpret their significance. These efforts reflect a universal quest to understand the universe and the place of humans within it. The myths and scientific achievements of these early cultures continue to inspire and inform contemporary astronomical research. By studying the past, we gain valuable insights into the enduring human fascination with the cosmos and the ways in which this fascination has driven scientific inquiry and cultural expression throughout history. As mankind moves forward in the exploration of space, new information and discoveries build upon the rich legacy of these early astronomers and storytellers, continuing the journey they began millennia ago.

BIBLIOGRAPHY

[Allen63] Allen, Richard Hinckley, *Star Names: Their Lore and Meaning*, Dover Publications, 1963.

[Aveni97] Aveni, Anthony F., *Stairways to the Stars: Skywatching in Three Great Ancient Cultures*, John Wiley & Sons, 1997.

[Evans98] Evans, James, *The History and Practice of Ancient Astronomy*, Oxford University Press, 1998.

[Geller11] Geller, Michael, *How to Find a Habitable Planet*, Princeton University Press, 2011.

[Krupp03] Krupp, Edwin C., *Echoes of the Ancient Skies: The Astronomy of Lost Civilizations*, Courier Corporation, 2003.

[Lankford97] Lankford, John, *History of Astronomy: An Encyclopedia*, Garland Publishing, 1997.

[McCluskey98] McCluskey, Stephen C., *Astronomies and Cultures in Early Medieval Europe*, Cambridge University Press, 1998.

[Morrison82] Morrison, Philip, and Phylis Morrison, *Powers of Ten: About the Relative Size of Things in the Universe*, Scientific American Library, 1982.

[O'ConnorRobertson] O'Connor, J. J., and E. F. Robertson, "The History of Ancient Astronomy," available online at https://mathshistory. st-andrews.ac.uk/HistTopics/Astronomy_in_ancient_times/.

[Ptolemy98] Ptolemy, *The Almagest*, Translated by G. J. Toomer, Princeton University Press, 1998.

[SuryaSiddhanta03] *Surya Siddhanta,* Translated by E. Burgess, Kessinger Publishing, 2003.

[Tyson17] Tyson, Neil deGrasse, *Astrophysics for People in a Hurry*, W.W. Norton & Company, 2017.

CHAPTER 2

THE DAWN OF THE SPACE AGE

THE SPACE RACE

The dawn of the Space Age marked a pivotal moment in human history characterized by a fierce competition between the United States and the Soviet Union. This period, known as the Space Race, saw both superpowers striving to achieve significant milestones in space exploration. The origins of the Space Race are deeply rooted in the Cold War, a period of intense geopolitical rivalry between the two nations. The launch of Sputnik 1 by the Soviet Union in 1957 not only marked the beginning of the Space Age but also intensified the competition, leading to an unprecedented era of scientific and technological advancements. This chapter explores the key achievements of the Space Race, including the launch of Sputnik, the first human in space Yuri Gagarin, and the historic Apollo 11 mission that landed humans on the moon. Discover the technological breakthroughs, political implications, and the profound impact of these achievements on the future of space exploration and international relations.

THE SPACE RACE: UNITED STATES VERSUS USSR

1. **Question**: What was the Space Race, and why did it begin?
   Answer: The Space Race was a period of intense competition between the United States and the Soviet Union, aiming to achieve superior spaceflight capabilities. It began in the context of the Cold War, with both superpowers seeking to demonstrate their technological and ideological dominance. The launch of Sputnik 1

by the USSR in 1957 marked the start of the Space Race, prompting the United States to accelerate its own space program. This rivalry drove rapid advancements in rocket technology and space exploration.

2. **Question**: What was the significance of Sputnik 1?
   Answer: Sputnik 1, launched by the Soviet Union on October 4, 1957, was the first artificial satellite to orbit the Earth. Its launch demonstrated the USSR's technological prowess and marked the beginning of the Space Age. Sputnik 1's success had significant political and military implications, as it showcased the capability to deploy satellites and potentially intercontinental ballistic missiles. The launch sparked widespread public interest and led to the establishment of NASA in the United States. (See Figure 2.1.)

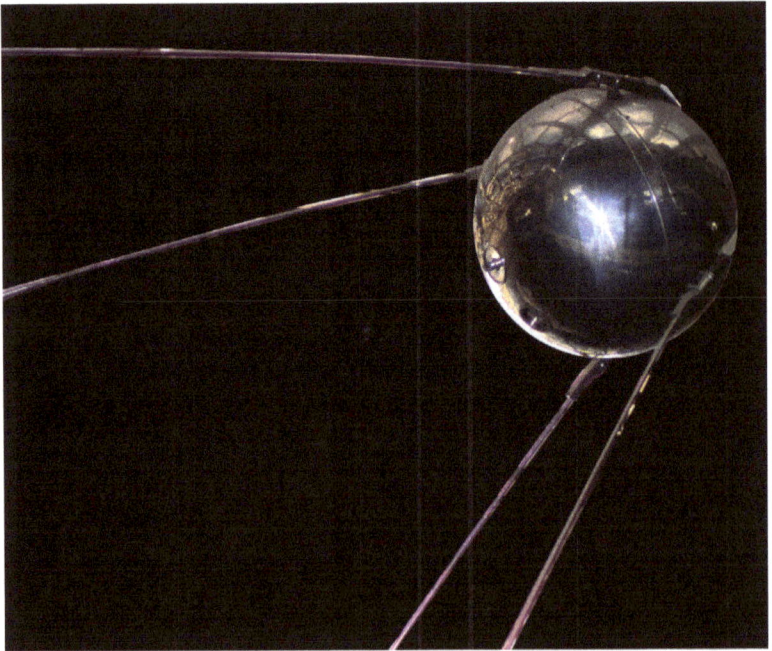

FIGURE 2.1 The launch of Sputnik 1 by the Soviet Union on October 4, 1957, marked the beginning of the Space Age and the start of the Space Race between the United States and the Soviet Union. Source: Wikimedia Commons.

3. **Question**: How did the United States respond to the launch of Sputnik 1?
   Answer: The United States responded to the launch of Sputnik 1 by accelerating its own space program. President Dwight D. Eisenhower established the National Aeronautics and Space

Administration (NASA) in 1958 to coordinate civilian space activities. The United States also increased funding for science and engineering education, emphasizing the need to compete with Soviet technological achievements. This response led to significant advancements in rocket technology and the eventual success of the Apollo program.

4. **Question**: What was the role of the Soviet space program in the early years of the Space Race?

   Answer: The Soviet space program, led by visionary engineers like Sergei Korolev, played a crucial role in the early years of the Space Race. The USSR achieved several firsts, including the launch of Sputnik 1, the first human in space, Yuri Gagarin, and the first woman in space, Valentina Tereshkova. These accomplishments established the Soviet Union as a formidable leader in space exploration and demonstrated their advanced capabilities in rocketry and spaceflight.

5. **Question**: Who was Sergei Korolev, and what was his contribution to the Soviet space program?

   Answer: Sergei Korolev was the chief engineer and mastermind behind the Soviet space program. He played a pivotal role in the development of the R-7 rocket, which launched Sputnik 1 and later missions. Korolev's leadership and engineering expertise were instrumental in achieving the early successes of the Soviet space program, including the first human spaceflight. His contributions laid the foundation for subsequent Soviet space achievements.

6. **Question**: How did the Space Race influence technological advancements?

   Answer: The Space Race spurred significant technological advancements in rocketry, satellite communications, materials science, and computing. Both superpowers invested heavily in research and development, leading to innovations that had far-reaching impacts beyond space exploration. Technologies developed during the Space Race contributed to advancements in various fields, including telecommunications, weather forecasting, and global positioning systems (GPS).

7. **Question**: What was the Mercury program, and what were its goals?

   Answer: The Mercury program was the first human spaceflight program of the United States, conducted by NASA between 1958 and 1963. Its primary goals were to place a human in orbit around the Earth, investigate human ability to function in space, and safely recover the astronaut and spacecraft. The program achieved several milestones, including the first American astronaut in space, Alan Shepard, and the first American in orbit, John Glenn.

8. **Question**: How did the Soviet Union achieve the first human spaceflight?
   Answer: The Soviet Union achieved the first human spaceflight with the launch of Vostok 1 on April 12, 1961, carrying cosmonaut Yuri Gagarin. The mission was a significant milestone in the Space Race, demonstrating the USSR's ability to send a human into space and return them safely to Earth. Gagarin's successful orbit of the Earth made him an international hero and solidified Soviet leadership in space exploration.

9. **Question**: What were the objectives of the Gemini program?
   Answer: The Gemini program was NASA's second human spaceflight program, running from 1961 to 1966. Its objectives included developing techniques for advanced space travel, such as long-duration missions, spacewalks (extravehicular activity), and orbital rendezvous and docking. These capabilities were crucial for the success of the subsequent Apollo program, which aimed to land humans on the moon.

10. **Question**: What impact did Yuri Gagarin's flight have on the Space Race?
    Answer: Yuri Gagarin's flight had a profound impact on the Space Race, propelling the Soviet Union to the forefront of space exploration. It intensified the competition between the United States and USSR, prompting the United States to accelerate its efforts to catch up. Gagarin's achievement was a major propaganda victory for the Soviet Union, showcasing their technological prowess and bolstering national pride.

KEY ACHIEVEMENTS: SPUTNIK, YURI GAGARIN, APOLLO 11

11. **Question**: What was the significance of the Vostok 1 mission?
    Answer: The Vostok 1 mission, which carried Yuri Gagarin into space on April 12, 1961, was significant for being the first human spaceflight. Gagarin's successful orbit of the Earth demonstrated the Soviet Union's advanced space capabilities and marked a major milestone in human space exploration. The mission's success had

profound geopolitical implications, influencing the course of the Space Race and inspiring future space missions.

12. **Question**: How did John Glenn's flight contribute to the Space Race?

    Answer: John Glenn's flight aboard Friendship 7 on February 20, 1962, made him the first American to orbit the Earth. This mission was a critical achievement for NASA and the United States, demonstrating their capability to send humans into orbit and recover them safely. Glenn's flight boosted American morale and confidence in their space program, narrowing the gap between the United States and USSR in the Space Race.

13. **Question**: What were the primary goals of the Apollo program?

    Answer: The primary goals of the Apollo program were to land humans on the moon and return them safely to Earth, as articulated by President John F. Kennedy in his 1961 speech. The program aimed to demonstrate American technological superiority and achieve a significant milestone in human space exploration. Additional goals included conducting scientific exploration of the lunar surface, developing human spaceflight capabilities, and establishing the United States as a leader in space exploration. (See Figure 2.2.)

FIGURE 2.2 President John F. Kennedy delivering his iconic "We choose to go to the moon" speech at Rice University on September 12, 1962. This moment set NASA's bold vision for space exploration. Credit: NASA

14. **Question**: How did the Apollo 11 mission achieve its objectives?
    Answer: The Apollo 11 mission, launched on July 16, 1969, achieved its objectives by successfully landing astronauts Neil Armstrong and Edwin "Buzz" Aldrin on the moon on July 20, 1969. The mission utilized the Saturn V rocket, the lunar module (Eagle), and the command module (Columbia) to carry out the complex tasks of lunar landing and return. Armstrong and Aldrin spent approximately twenty-four hours on the lunar surface, conducting experiments and collecting samples, while Michael Collins orbited the moon in the Command Module. The mission concluded with a safe return to Earth on July 24, 1969.

15. **Question**: What were the scientific achievements of the Apollo 11 mission?
    Answer: The Apollo 11 mission's scientific achievements included the collection of 47.5 pounds of lunar rock and soil samples, the deployment of scientific instruments such as the Passive Seismic Experiment and the Lunar Ranging Retroreflector, and extensive photographic documentation of the lunar surface. These contributions provided valuable data on the moon's composition, structure, and geological history, advancing our understanding of the moon and the early solar system.

16. **Question**: How did the success of Apollo 11 impact the Space Race?
    Answer: The success of Apollo 11 marked the culmination of the Space Race, securing a decisive victory for the United States. The successful moon landing demonstrated American technological prowess and fulfilled President Kennedy's goal of landing a man on the moon before the end of the 1960s. This achievement boosted national pride, inspired future space exploration, and solidified NASA's reputation as a leader in space science and technology.

17. **Question**: What role did the Saturn V rocket play in the Apollo program?
    Answer: The Saturn V rocket, designed by a team led by Wernher von Braun, played a crucial role in the Apollo program by providing the necessary thrust to send astronauts to the moon. As the most powerful rocket ever built, the Saturn V stood 363 feet tall and generated 7.5 million pounds of thrust. Its three-stage design allowed it to deliver the Apollo spacecraft into lunar orbit, enabling the historic moon landings. The reliability and success of the Saturn V were instrumental in achieving the Apollo program's goals.

18. **Question**: How did the Apollo program influence subsequent space exploration?

    Answer: The Apollo program influenced subsequent space exploration by demonstrating the feasibility of human space travel beyond Earth orbit, advancing spacecraft technology, and fostering international collaboration in space science. The scientific discoveries and technological innovations developed during Apollo paved the way for future missions, including the Space Shuttle program, the International Space Station (ISS), and plans for Mars exploration. Apollo's legacy continues to inspire and inform contemporary efforts in space exploration.

19. **Question**: What was the first satellite launched by China, and why was it significant?

    Answer: China's first satellite, Dong Fang Hong 1, was launched on April 24, 1970. It marked China's entry into the Space Age and demonstrated its capability to develop and deploy space technology independently.

20. **Question**: What challenges did the Apollo 11 mission overcome to achieve success?

    Answer: The Apollo 11 mission overcame numerous challenges, including the development of reliable spacecraft systems, ensuring astronaut safety, navigating the complex lunar landing, and coordinating a successful return to Earth. Technical challenges included designing the lunar module for a safe descent and ascent, ensuring the Command Module's reentry capability, and managing limited fuel reserves during the landing. The mission's success was a testament to the ingenuity, dedication, and collaboration of thousands of engineers, scientists, and astronauts.

21. **Question**: How did the international community react to the Apollo 11 moon landing?

    Answer: The international community reacted to the Apollo 11 moon landing with awe and admiration. The historic event was broadcast live around the world, uniting people in a shared moment of human achievement. Leaders from various countries congratulated the United States, recognizing the significance of the moon landing for all humanity. The mission was seen as a triumph of scientific and technological progress, transcending national boundaries and inspiring future generations to explore the cosmos. (See Figure 2.3.)

FIGURE 2.3 The launch of Apollo 11 on July 16, 1969, from Kennedy Space Center, which carried astronauts Neil Armstrong, Buzz Aldrin, and Michael Collins to the moon. Source: Wikimedia Commons.

22. **Question**: What were the long-term effects of the Space Race on scientific research and development?

    Answer: The long-term effects of the Space Race on scientific research and development include significant advancements in aerospace technology, telecommunications, computer science, and materials engineering. The investments and innovations driven by the Space Race laid the groundwork for modern satellite technology, GPS systems, and advancements in microelectronics. The collaborative efforts and competition between the United States and

USSR accelerated the pace of scientific discovery and technological progress, benefiting various fields beyond space exploration.

23. **Question**: How did the Space Race shape public perception of space exploration?
    Answer: The Space Race shaped public perception of space exploration by capturing the imagination and interest of people worldwide. The dramatic achievements and milestones, such as the first human spaceflight and the moon landing, highlighted the potential of space exploration and its impact on human progress. The Space Race also emphasized the importance of scientific and technological education, inspiring many to pursue careers in STEM fields. Public support for space programs grew, fostering a sense of wonder and aspiration for future space endeavors.

24. **Question**: How did India's first satellite, Aryabhata, contribute to space exploration?
    Answer: Launched on April 19, 1975, Aryabhata was India's first satellite. It marked a significant milestone in India's space program, enabling advancements in space technology and scientific research.

25. **Question**: What was the legacy of Yuri Gagarin's historic flight?
    Answer: Yuri Gagarin's historic flight left a lasting legacy as a symbol of human achievement and the possibilities of space exploration. Gagarin's successful orbit of the Earth demonstrated the feasibility of human spaceflight and inspired future generations of astronauts and space enthusiasts. His flight also highlighted the importance of international cooperation and competition in advancing space science. Gagarin remains a celebrated figure in the history of space exploration, commemorated for his pioneering spirit and contributions to human spaceflight.

26. **Question**: What was the significance of the lunar module in the Apollo missions?
    Answer: The lunar module was significant in the Apollo missions because it was the spacecraft designed to land astronauts on the moon and return them to the Command Module in lunar orbit. Developed by Grumman Aerospace, the lunar module had a two-stage design: the descent stage for landing on the moon and the ascent stage for returning to lunar orbit. The lunar module's innovative design allowed it to operate in the vacuum of space and withstand the harsh lunar environment. Its successful deployment

and operation were critical to the success of the Apollo moon landings, enabling astronauts to explore the lunar surface and conduct scientific experiments.

27. **Question**: What role did the media play in shaping public perception of the Space Race?

    Answer: The media played a crucial role in shaping public perception of the Space Race by providing extensive coverage of space missions, astronaut achievements, and technological advancements. Television broadcasts, newspaper articles, and radio reports brought the excitement and drama of space exploration into homes around the world. Iconic moments, such as the launch of Sputnik, Yuri Gagarin's flight, and the Apollo 11 moon landing, were broadcast live, capturing the imagination of millions. The media's portrayal of the Space Race as a symbol of national pride and technological progress helped to garner public support for space programs and inspire future generations. (See Figure 2.4.)

FIGURE 2.4 Yuri Gagarin became the first human to journey into outer space and orbit the Earth on April 12, 1961. This milestone was a significant victory for the Soviet space program. Source: Wikimedia Commons.

28. **Question**: How did the Space Race contribute to the advancement of computer technology?
    Answer: The Space Race contributed to the advancement of computer technology by driving the development of more powerful and reliable computing systems for space missions. Early space missions required sophisticated onboard computers for navigation, control, and data processing. The need for miniaturized, robust, and efficient computers led to innovations in hardware and software design. NASA's use of integrated circuits in the Apollo Guidance Computer, for example, paved the way for the broader adoption of microelectronics in various industries. The advancements in computer technology during the Space Race had far-reaching implications, accelerating the growth of the digital age.

29. **Question**: What was the significance of the Shenzhou 5 mission for China?
    Answer: The Shenzhou 5 mission, launched on October 15, 2003, carried Yang Liwei, China's first astronaut, into space. This mission established China as the third country capable of independently sending humans into orbit.

30. **Question**: What was the significance of the first spacewalk, and who conducted it?
    Answer: The first spacewalk, or extravehicular activity (EVA), was significant because it demonstrated that astronauts could leave their spacecraft and perform tasks in the vacuum of space. This milestone was achieved by Soviet cosmonaut Alexei Leonov on March 18, 1965, during the Voskhod 2 mission. Leonov's spacewalk lasted approximately twelve minutes, during which he maneuvered outside the spacecraft while tethered to it. The success of this EVA proved that human activities outside a spacecraft were feasible, paving the way for future spacewalks, construction of space stations, and lunar exploration.

31. **Question**: How did the Space Race influence international policies and treaties regarding space exploration?
    Answer: The Space Race influenced international policies and treaties by highlighting the need for regulations to ensure the peaceful use of outer space. Key agreements, such as the Outer Space Treaty of 1967, established principles for the exploration and use of space, including the prohibition of placing nuclear weapons in orbit and the declaration that space should be used for the benefit

of all humankind. The treaty also emphasized that celestial bodies, such as the moon, should not be subject to national appropriation. These policies and treaties helped to promote international cooperation and prevent conflicts in space.

32. **Question**: What was the impact of the Space Race on the Cold War?
Answer: The Space Race impacted the Cold War by serving as a nonmilitary arena for competition between the United States and USSR, allowing both superpowers to demonstrate their technological and ideological superiority. The achievements in space exploration were seen as indicators of national strength and scientific prowess. The Space Race also intensified the rivalry, leading to increased investments in aerospace research and development. However, it also provided opportunities for diplomatic engagement and collaboration, such as the Apollo-Soyuz Test Project, which helped to ease tensions and promote mutual understanding.

33. **Question**: How did the development of rocket technology progress during the Space Race?
Answer: The development of rocket technology progressed rapidly during the Space Race, with both the United States and USSR achieving significant advancements in propulsion, guidance, and launch systems. The Soviet Union's R-7 rocket, which launched Sputnik and Vostok missions, was a pioneering design that demonstrated the feasibility of multistage rockets. In the United States, the development of the Saturn V rocket by NASA enabled the Apollo missions to reach the moon. These advancements were driven by the need for powerful, reliable rockets capable of carrying heavy payloads and achieving precise orbits.

34. **Question**: What role did space probes play in the Space Race?
Answer: Space probes played a vital role in the Space Race by providing valuable data on the moon, planets, and other celestial bodies. Both the United States and USSR launched a series of robotic missions to explore the solar system. The Soviet Luna program achieved the first successful impact on the moon and returned the first lunar soil samples. The American Ranger, Surveyor, and Lunar Orbiter programs provided detailed images and data that informed the planning of manned lunar landings. These robotic missions expanded our understanding of the solar system and demonstrated the potential of unmanned exploration.

35. **Question**: How did the Space Race influence space law and policy?
    Answer: The Space Race influenced space law and policy by highlighting the need for international agreements to govern the exploration and use of outer space. The Outer Space Treaty of 1967 established principles for the peaceful use of space, including the prohibition of nuclear weapons in orbit and the declaration that space is free for exploration by all nations. The treaty also emphasized that space activities should benefit all humankind and that celestial bodies should not be subject to national appropriation. These principles continue to guide space exploration and international cooperation in space activities.

36. **Question**: What were the main scientific objectives of the Apollo lunar missions?
    Answer: The main scientific objectives of the Apollo lunar missions included studying the moon's geology, collecting rock and soil samples, and deploying scientific instruments on the lunar surface. The missions aimed to understand the moon's composition, structure, and history, providing insights into the early solar system. Instruments such as seismometers, magnetometers, and retroreflectors were used to measure moonquakes, magnetic fields, and distances between the Earth and moon. The samples collected by astronauts were analyzed to determine the moon's mineralogy and age, contributing to our knowledge of planetary formation and evolution.

37. **Question**: What achievements did the Indian Space Research Organisation (ISRO) accomplish with the Polar Satellite Launch Vehicle (PSLV) program?
    Answer: The PSLV program has been a cornerstone of ISRO's success, with over fifty successful launches. Notably, it placed the Mars Orbiter Mission (Mangalyaan) into Mars orbit, making India the first Asian nation to reach Mars and the fourth space agency to reach Mars.

38. **Question**: How did the Space Race set the stage for future space exploration?
    Answer: The Space Race set the stage for future space exploration by establishing the technological and scientific foundations necessary for advanced missions. The achievements of the Space Race, including human spaceflight, lunar landings, and robotic exploration, demonstrated the feasibility of ambitious space endeavors.

The experience gained, technologies developed, and international collaborations initiated during this period paved the way for subsequent programs such as the Space Shuttle, the ISS, and planned missions to Mars. The legacy of the Space Race continues to inspire and drive the pursuit of new frontiers in space exploration.

39. **Question**: How did the Space Race contribute to the development of satellite technology?
    Answer: The Space Race accelerated the development of satellite technology, leading to significant advancements in communication, weather forecasting, and Earth observation. The successful launch of early satellites like Sputnik 1 and Explorer 1 demonstrated the feasibility of deploying artificial satellites in orbit. Subsequent missions, such as the TIROS weather satellites and the Telstar communication satellites, showcased the practical applications of satellite technology. The innovations and investments made during the Space Race laid the foundation for the modern satellite industry, which continues to play a critical role in global communications and scientific research.

40. **Question**: How did the Space Race impact education in the United States?
    Answer: The Space Race significantly impacted education in the United States by promoting increased funding and emphasis on science, technology, engineering, and mathematics (STEM) education. The launch of Sputnik 1 and subsequent Soviet achievements in space highlighted the need for the United States to strengthen its scientific and technical workforce. In response, the US government established programs to support STEM education, including scholarships, grants, and the creation of specialized schools and research institutions. This focus on education contributed to the development of a highly skilled workforce that supported the nation's space program and technological advancements.

41. **Question**: How did the Space Race influence popular culture?
    Answer: The Space Race influenced popular culture by inspiring a wide range of media, including films, television shows, books, and music. The dramatic achievements and futuristic vision of space exploration captured the public's imagination, leading to the creation of iconic works such as "Star Trek," "2001: A Space Odyssey," and "The Right Stuff." These cultural artifacts reflected society's fascination with space and the possibilities of human achievement

beyond Earth. The Space Race also contributed to the portrayal of astronauts as heroes and role models, influencing the aspirations of future generations.

42. **Question**: What role did women play in the Space Race?
    Answer: Women played crucial roles in the Space Race, contributing as engineers, mathematicians, scientists, and astronauts. In the United States, women like Katherine Johnson, Dorothy Vaughan, and Mary Jackson made significant contributions to NASA's calculations and mission planning. In the Soviet Union, Valentina Tereshkova became the first woman to travel to space in 1963. Despite facing gender-based challenges and discrimination, many women made lasting impacts on the success of space missions and advancements in space science.

43. **Question**: What was the Apollo-Soyuz Test Project, and why was it significant?
    Answer: The Apollo-Soyuz Test Project, conducted in 1975, was the first international manned spaceflight mission, where American and Soviet spacecraft docked in orbit. This mission was significant because it marked the end of the Space Race and the beginning of international cooperation in space exploration. The project demonstrated that the U nited States and USSR could work together despite political tensions, setting a precedent for future collaborative efforts in space. The mission included joint scientific experiments and symbolic gestures of goodwill, fostering a spirit of collaboration that paved the way for projects like the ISS.

44. **Question**: How did the Space Race affect the development of international space agencies?
    Answer: The Space Race impacted the development of international space agencies by demonstrating the importance and benefits of space exploration, inspiring other nations to establish their own space programs. Countries such as France, Japan, India, and China began to develop their space capabilities, leading to the creation of agencies like the European Space Agency (ESA) and the Indian Space Research Organisation (ISRO). These agencies have since contributed to global space exploration efforts, conducting scientific missions, launching satellites, and participating in international collaborations such as the ISS.

45. **Question**: What technological innovations from the Space Race have been adapted for everyday use?
    Answer: Several technological innovations from the Space Race have been adapted for everyday use, including advancements in materials science, telecommunications, and medical technology. For example, the development of lightweight, durable materials for spacecraft has led to improvements in consumer products such as athletic shoes and firefighting gear. Satellite technology developed during the Space Race has revolutionized global communications, enabling satellite TV, GPS navigation, and weather forecasting. Medical innovations, such as digital imaging and remote monitoring, have also benefited from technologies originally designed for space missions.

46. **Question**: How did the Space Race contribute to the understanding of human physiology in space?
    Answer: The Space Race contributed to the understanding of human physiology in space by providing valuable data on the effects of microgravity and space travel on the human body. Astronauts participating in missions conducted experiments and recorded their physical responses to weightlessness, radiation exposure, and isolation. These studies revealed changes in bone density, muscle atrophy, fluid distribution, and cardiovascular function. The findings have informed the development of countermeasures to mitigate these effects, such as exercise protocols and medical interventions, ensuring the health and safety of astronauts on long-duration missions.

47. **Question**: What was the role of the Jet Propulsion Laboratory (JPL) during the Space Race?
    Answer: The JPL played a crucial role during the Space Race by developing and managing various robotic space missions. The JPL's contributions included the successful launch and operation of the Explorer 1 satellite, America's first artificial satellite, and the development of interplanetary probes such as Mariner, Ranger, and Surveyor. These missions provided valuable data on the moon, Mars, Venus, and other celestial bodies, supporting NASA's broader goals of space exploration. The JPL's expertise in engineering and mission management was instrumental in advancing the United States' capabilities in space science and exploration.

48. **Question**: How did the Space Race address the challenge of reentry and landing?

    Answer: The Space Race addressed the challenge of reentry and landing by developing heat shields, reentry capsules, and parachute systems to protect astronauts during their return to Earth. Spacecraft reentering the Earth's atmosphere encounter extreme heat and friction, requiring robust heat shields to prevent burning up. The Mercury, Gemini, and Apollo capsules were designed with ablative heat shields that dissipated heat through material erosion. Parachute systems were used to slow the descent of the capsules, ensuring a safe landing. The development and testing of these technologies were critical to the success of human spaceflight missions. (See Figure 2.5.)

FIGURE 2.5 Astronaut Buzz Aldrin stands on the moon next to the American flag during the Apollo 11 mission, showcasing the historic first manned lunar landing.
Source: Wikimedia Commons

49. **Question**: How did the international community react to Yuri Gagarin's flight?

    Answer: The international community reacted to Yuri Gagarin's flight with a mixture of awe, admiration, and competitive determination. Gagarin's successful orbit was a monumental achievement that showcased the Soviet Union's technological capabilities and inspired people worldwide. Leaders from various countries congratulated the Soviet Union, recognizing the historic significance of the flight. At the same time, Gagarin's achievement intensified the competition within the Space Race, prompting the United States to accelerate its efforts to achieve its own milestones in space exploration.

50. **Question**: What scientific discoveries were made possible by the Apollo lunar missions?

    Answer: The Apollo lunar missions made several scientific discoveries, including the identification of lunar rock types, evidence of volcanic activity, and insights into the moon's thermal history. The collection of rock and soil samples revealed the presence of basalt and anorthosite, providing evidence of past volcanic activity and the moon's geological evolution. Instruments left on the lunar surface, such as seismometers, detected moonquakes, which offered clues about the moon's internal structure. These discoveries advanced our understanding of the moon's formation and its relationship to Earth and the broader solar system.

51. **Question**: What role did the Space Race play in the development of new materials and manufacturing techniques?

    Answer: The Space Race played a significant role in the development of new materials and manufacturing techniques by driving the need for lightweight, durable, and high-performance materials for spacecraft and equipment. Innovations included the development of advanced composites, high-strength alloys, and thermal protection materials. These materials were essential for building rockets, space capsules, and other components that could withstand the harsh conditions of space travel. The advancements in materials science and manufacturing techniques developed during the Space Race have since been applied to various industries, including aviation, automotive, and consumer electronics.

52. **Question**: How did the Space Race influence the establishment of space policies and governance?

    Answer: The Space Race influenced the establishment of space policies and governance by highlighting the need for international cooperation and regulation of space activities. Key milestones included the signing of the Outer Space Treaty in 1967, which established principles for the peaceful use of outer space and prohibited the deployment of nuclear weapons in space. The treaty also emphasized that space exploration should benefit all humankind and that celestial bodies should not be subject to national appropriation. These policies laid the foundation for international space law and continue to guide the governance of space activities today.

53. **Question**: What was the significance of the Ranger missions in the context of the Space Race?

    Answer: The Ranger missions were significant in the context of the Space Race because they provided the first close-up images of the moon's surface, which were crucial for planning the Apollo lunar landings. The Ranger program, conducted by NASA, consisted of a series of missions designed to capture high-resolution images of the moon. The successful Ranger missions, particularly Ranger 7, 8, and 9, provided detailed photographs that helped scientists and engineers select landing sites and understand the lunar terrain. These missions demonstrated the capability of robotic exploration and contributed to the overall success of the Apollo program.

54. **Question**: How did the Space Race contribute to the development of international space collaborations?

    Answer: The Space Race contributed to the development of international space collaborations by demonstrating the benefits of shared scientific and technological efforts. While initially characterized by intense competition, the Space Race eventually led to cooperative projects such as the Apollo-Soyuz Test Project in 1975, where American and Soviet spacecraft docked in orbit. This collaboration set the stage for future international partnerships, such as the development of the ISS, where multiple countries work together to conduct research and explore space. The legacy of the Space Race includes a spirit of international cooperation that continues to drive advancements in space exploration today.

The dawn of the Space Age, driven by the intense competition between the United States and the Soviet Union, led to monumental achievements that forever changed our understanding of space and our place within it. The Space Race catalyzed rapid advancements in rocketry, satellite technology, and human spaceflight, culminating in the iconic Apollo 11 moon landing. These achievements were not merely technological milestones; they were powerful symbols of national pride and human ingenuity. The legacy of the Space Race extends beyond the Cold War, influencing international space policy, fostering global cooperation, and inspiring generations of scientists, engineers, and explorers. As we look back on this transformative period, it is clear that the Space Race laid the foundation for modern space exploration and established the framework for international partnerships in space endeavors. The spirit of competition and collaboration that characterized the Space Race continues to drive humanity's quest to explore the final frontier, reminding us of what we can achieve when we reach for the stars.

REFERENCES

[Cadbury06] Cadbury, Deborah, *Space Race: The Epic Battle Between America and the Soviet Union for Dominion of Space*, HarperCollins, 2006.

[Chaikin94] Chaikin, Andrew, *A Man on the Moon: The Voyages of the Apollo Astronauts*, Penguin Books, 1994.

[Harford97] Harford, James, *Korolev: How One Man Masterminded the Soviet Drive to Beat America to the Moon*, John Wiley & Sons, 1997.

[Krige13] Krige, John, *NASA in the World: Fifty Years of International Collaboration in Space*, Palgrave Macmillan, 2013.

[Logsdon10] Logsdon, John M., *John F. Kennedy and the Race to the Moon*, Palgrave Macmillan, 2010.

[Neufeld07] Neufeld, Michael J., *Von Braun: Dreamer of Space, Engineer of War*, Alfred A. Knopf, 2007.

[Siddiqi00] Siddiqi, Asif A., *Challenge to Apollo: The Soviet Union and the Space Race, 1945–1974*, NASA History Series SP-2000-4408, Military Book Shop.

CHAPTER 3

ROBOTIC EXPLORERS BEYOND EARTH

INTRODUCTION

Robotic explorers have played a crucial role in advancing our understanding of the solar system and beyond. This chapter focuses on the remarkable achievements of Mars rovers and landers, the pioneering Voyager and New Horizons missions, and the international collaborations that have pushed the boundaries of robotic space exploration. From the early missions that provided humanity with the first close-up images of our neighboring planets, to the sophisticated rovers and probes that continue to send back data from the far reaches of our solar system, robotic explorers have been at the forefront of space science. Behind each of these missions is a myriad of technological innovations, scientific discoveries, and collaborative efforts that made them possible to achieve. This chapter highlights how these robotic pioneers have not only expanded our knowledge of distant worlds but also laid the groundwork for future human exploration.

MARS ROVERS AND LANDERS

1. **Question**: What was the first successful Mars lander, and what did it achieve?

    Answer: The first successful Mars lander was Viking 1, launched by NASA in 1975. Viking 1 became the first spacecraft to send back high-resolution images of the Martian surface and conduct scientific experiments on Mars. It analyzed soil samples, searched for signs of life, and provided valuable data on Martian weather and geology. Viking 1's success paved the way for future Mars missions.

2. **Question**: How did the Pathfinder mission contribute to Mars exploration?
   Answer: The Pathfinder mission, which landed on Mars in 1997, was significant for deploying the first successful rover, Sojourner. This mission demonstrated new technologies for entry, descent, and landing, including an airbag-cushioned landing system. Sojourner conducted chemical analyses of rocks and soil, providing insights into the planet's composition and surface conditions.

3. **Question**: What were the objectives of the Spirit and Opportunity rovers?
   Answer: The objectives of the Spirit and Opportunity rovers, part of NASA's Mars Exploration Rover (MER) mission launched in 2003, were to search for and characterize a wide range of rocks and soils that might hold clues to past water activity on Mars. Both rovers provided critical evidence of ancient water flows, diverse mineral compositions, and dynamic environmental conditions on Mars.

4. **Question**: How did the Curiosity rover enhance our understanding of Mars?
   Answer: The Curiosity rover, part of NASA's Mars Science Laboratory mission, landed on Mars in 2012. It was designed to assess the planet's habitability by studying its climate, geology, and potential for past life. Curiosity's discoveries include the detection of complex organic molecules, signs of ancient lakes and streams, and fluctuating levels of methane in the atmosphere.

5. **Question**: What innovations did the Perseverance rover bring to Mars exploration?
   Answer: Launched in 2020, NASA's Perseverance rover introduced several innovations, including the first attempt to produce oxygen from Martian CO_2 through the MOXIE experiment, the Ingenuity helicopter for aerial reconnaissance, and advanced caching systems for sample collection. Its mission is to search for signs of ancient microbial life and collect samples for future return to Earth. (See Figure 3.1.)

FIGURE 3.1 An iconic image of the Perseverance rover alongside the Ingenuity helicopter on Mars. This photo showcases the innovative technology of the Mars 2020 mission, aimed at exploring Jezero Crater for signs of ancient microbial life. Source: Wikimedia Commons.

6. Question: What was the significance of the Sojourner rover?
Answer: The Sojourner rover, part of the Mars Pathfinder mission, was significant as the first rover to operate on Mars. It demonstrated the feasibility of mobile exploration and provided valuable data on Martian soil mechanics, dust, and weather conditions. Its successful operation paved the way for future rovers like Spirit, Opportunity, Curiosity, and Perseverance.

7. Question: What discoveries did the Phoenix lander make on Mars?
Answer: The Phoenix lander, which arrived on Mars in 2008, discovered water-ice just below the Martian surface. It analyzed soil samples and found evidence of perchlorates, indicating the presence of liquid water in the past. Phoenix's findings contributed to our understanding of the Martian polar environment and its potential for supporting life.

8. Question: How did the InSight mission contribute to our knowledge of Mars's interior?
Answer: The InSight mission, which landed on Mars in 2018, provided valuable data on Mars's interior structure. It deployed a seismometer to detect Marsquakes and a heat flow probe to measure the planet's thermal properties. InSight's findings have improved our understanding of Mars's geologic activity and its interior composition.

9. **Question**: What was the role of the Mars Global Surveyor?
   Answer: The Mars Global Surveyor, launched in 1996, mapped the entire Martian surface with high-resolution imaging and collected data on the planet's topography, gravity, and magnetic field. Its observations provided critical insights into Mars's climate history, surface processes, and potential for water.

10. **Question**: How did the Mars Odyssey mission enhance our understanding of Mars?
    Answer: The Mars Odyssey mission, launched in 2001, detected significant amounts of hydrogen below the Martian surface, suggesting the presence of water-ice. It also mapped the planet's mineral composition and monitored radiation levels, providing valuable data for future human missions to Mars.

THE VOYAGER AND NEW HORIZONS MISSIONS

11. **Question**: What were the primary goals of the Voyager missions?
    Answer: The primary goals of the Voyager missions, launched by NASA in 1977, were to explore the outer planets and their moons. Voyager 1 and Voyager 2 provided detailed images and data on Jupiter, Saturn, Uranus, and Neptune, making significant discoveries such as active volcanoes on Io and geysers on Triton.

12. **Question**: How did Voyager 1 contribute to our understanding of the interstellar medium?
    Answer: Voyager 1, now the farthest human-made object from Earth, entered interstellar space in 2012. It provided the first direct measurements of the interstellar medium, including data on cosmic rays, plasma waves, and magnetic fields, giving scientists unprecedented insights into the environment beyond our solar system's influence. (See to Figure 3.2.)

FIGURE 3.2 An image of the Voyager spacecraft during vibration testing, illustrating the sophisticated design and technology used in the mission. Source: Wikimedia Commons.

13. **Question**: What was the significance of the Golden Record on the Voyager spacecraft?
    Answer: The Golden Record, included on both Voyager spacecraft, is a phonograph record containing sounds and images selected to portray the diversity of life and culture on Earth. It was intended as a message to any extraterrestrial civilizations that might encounter the spacecraft, representing humanity's desire to communicate with the cosmos.

14. **Question**: What were the key findings of the New Horizons mission at Pluto?
    Answer: The New Horizons mission, which conducted a historic flyby of Pluto in 2015, revealed a diverse landscape with mountains of water ice, vast plains of nitrogen ice, and evidence of geological activity. It transformed our understanding of Pluto, highlighting its complexity and dynamic nature.

15. **Question**: How did New Horizons contribute to the study of the Kuiper Belt?
    Answer: After its Pluto flyby, New Horizons continued its journey to the Kuiper Belt. In 2019, it conducted a flyby of Arrokoth, providing detailed images and data on this distant object. The mission's observations revealed a contact binary shape and provided insights into the formation and evolution of Kuiper Belt objects.

16. **Question**: What discoveries did Voyager 2 make during its mission?
    Answer: Voyager 2 made significant discoveries during its mission, including detailed observations of Jupiter's Great Red Spot, Saturn's rings, Uranus's tilted magnetic field, and Neptune's active weather systems. It also provided the first close-up images of these planets and their moons, greatly expanding our knowledge of the outer solar system.

17. **Question**: How did the Voyager missions change our understanding of the solar system?
    Answer: The Voyager missions transformed our understanding of the solar system by providing detailed observations of the outer planets and their moons, revealing complex and dynamic worlds. Their discoveries included volcanic activity on Io, subsurface oceans on Europa, and the diverse ring systems of Saturn and Uranus.

18. **Question**: What is the significance of the "Pale Blue Dot" image taken by Voyager 1?
    Answer: The "Pale Blue Dot" image, taken by Voyager 1 in 1990, shows Earth as a tiny speck in the vastness of space. This iconic photograph highlights the fragility and isolation of our planet, emphasizing the importance of preserving our environment and fostering a sense of unity among humanity.

19. **Question**: How have the Voyager spacecraft continued to contribute to science after their primary missions?
    Answer: The Voyager spacecraft have continued to contribute to science by providing data on the outer edges of the solar system and the interstellar medium. Their instruments are still operational, sending back valuable information about cosmic rays, solar winds, and the heliopause, the boundary where the solar wind meets interstellar space.

20. **Question**: What technological innovations were used in the New Horizons mission?
    Answer: The New Horizons mission utilized several technological innovations, including advanced imaging systems, a compact and lightweight design, and a nuclear power source for long-duration missions. Its instruments provided high-resolution images and spectral data, enabling detailed studies of Pluto and its moons.

INTERNATIONAL COLLABORATIONS IN ROBOTIC SPACE MISSIONS

21. **Question**: What is the significance of the European Space Agency's Mars Express mission?
    Answer: Launched in 2003, the European Space Agency's Mars Express mission has provided valuable data on the Martian atmosphere, surface, and subsurface. One of its significant achievements is the detection of subsurface water ice and hydrated minerals, suggesting that liquid water may exist below the surface.

22. **Question**: How did the collaboration between NASA and ESA enhance the Cassini-Huygens mission?
    Answer: The Cassini-Huygens mission, a collaboration between NASA, ESA, and the Italian Space Agency, provided unprecedented insights into Saturn and its moons. Cassini orbited Saturn for thirteen years, studying its rings, atmosphere, and magnetic field, while the Huygens probe landed on Titan, revealing its surface and atmosphere.

23. **Question**: What role did the Rosetta mission play in advancing comet science?

    Answer: The European Space Agency's Rosetta mission, launched in 2004, was the first to orbit and land a probe on a comet. Rosetta studied Comet 67P/Churyumov-Gerasimenko, providing detailed data on its composition, structure, and activity, advancing our understanding of these ancient, primordial bodies.

24. **Question**: How have international collaborations contributed to lunar exploration?

    Answer: International collaborations have significantly advanced lunar exploration, with missions such as the Lunar Reconnaissance Orbiter (LRO) and Chandrayaan-1. The LRO has mapped the moon's surface in high detail, while Chandrayaan-1 discovered water molecules on the moon's surface. (See Figure 3.3.)

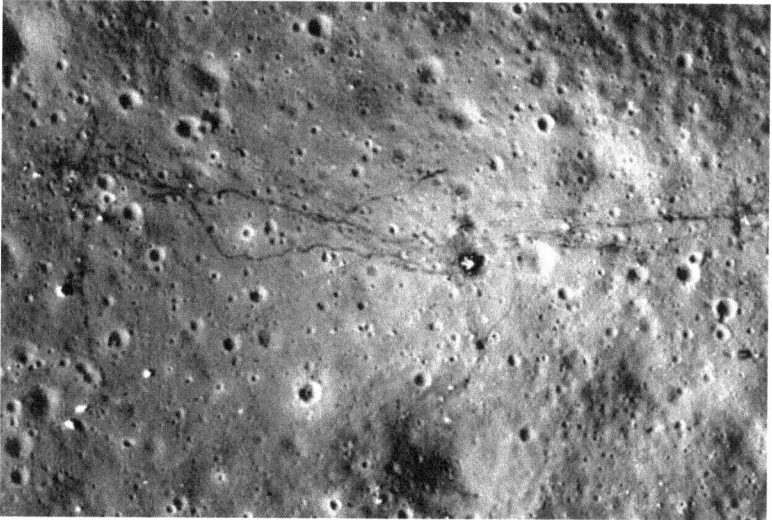

FIGURE 3.3 This image from the Lunar Reconnaissance Orbiter Camera (LROC) shows the Apollo landing sites and the tracks left by astronauts on the Moon. The LROC has been instrumental in mapping the lunar surface and supporting various international lunar exploration missions. Source: Wikimedia Commons.

25. **Question**: What impact has the International Space Station (ISS) had on robotic space missions?

    Answer: The ISS has served as a platform for numerous robotic space missions and experiments. It has hosted a variety of robotic systems, such as the Canadarm2, which assists with spacecraft docking and maintenance tasks, and has supported experiments involving robotic explorers.

26. **Question**: How did the Japan Aerospace Exploration Agency (JAXA) contribute to asteroid exploration?

    Answer: JAXA's Hayabusa and Hayabusa2 missions were significant contributions to asteroid exploration. Hayabusa was the first mission to return samples from an asteroid, Itokawa, while Hayabusa2 returned samples from the asteroid Ryugu, providing valuable data on the composition and structure of asteroids.

27. **Question**: What is the significance of the ExoMars program?

    Answer: The ExoMars program, a collaboration between ESA and Roscosmos, aims to search for signs of past or present life on Mars. The program includes the Trace Gas Orbiter, which studies the Martian atmosphere, and the upcoming Rosalind Franklin rover, which will analyze subsurface samples.

28. **Question**: How has the collaboration between NASA and the Canadian Space Agency (CSA) advanced space robotics?

    Answer: The collaboration between NASA and the CSA has advanced space robotics through the development of robotic arms like the Canadarm and Canadarm2. These systems have been instrumental in deploying satellites, conducting repairs, and assembling the ISS.

29. **Question**: How have international collaborations enhanced Mars exploration?

    Answer: International collaborations have enhanced Mars exploration by combining the expertise and resources of multiple space agencies. Missions like the Mars Science Laboratory (Curiosity), ExoMars, and the Mars Reconnaissance Orbiter have benefited from contributions by NASA, ESA, Roscosmos, and other agencies.

30. **Question**: What are the key achievements of China's Chang'e lunar missions?

    Answer: China's Chang'e lunar missions have achieved significant milestones, including the first soft landing on the Moon's far side by Chang'e 4 and the successful return of lunar samples by Chang'e 5. These missions have advanced our understanding of the Moon's geology and potential resources.

31. **Question**: What role did the Hubble Space Telescope play in international space collaboration?

    Answer: The Hubble Space Telescope, a collaboration between NASA and ESA, has provided unprecedented views of the universe. Its discoveries, such as the acceleration of the universe's expansion and detailed images of distant galaxies, have been accessible to scientists worldwide.

32. **Question**: How has the collaboration between NASA and private companies advanced robotic space exploration?
    Answer: Collaboration between NASA and private companies has advanced robotic space exploration by leveraging commercial innovation and efficiency. Programs like the Commercial Lunar Payload Services (CLPS) have accelerated the development and deployment of robotic landers, rovers, and payloads.

33. **Question**: What is the significance of the Artemis program for robotic exploration?
    Answer: The Artemis program, led by NASA, aims to return humans to the moon and establish a sustainable presence. It involves robotic exploration, including lunar rovers and scientific instruments, to conduct reconnaissance, study lunar resources, and test technologies for human exploration.

34. **Question**: How have advancements in AI and robotics influenced space exploration?
    Answer: Advancements in AI and robotics have improved the autonomy and capabilities of robotic missions. AI enables spacecraft and rovers to make real-time decisions, optimize navigation, and conduct complex scientific analyses without direct human intervention.

35. **Question**: What impact have international space organizations had on robotic missions?
    Answer: International space organizations, such as the United Nations Office for Outer Space Affairs (UNOOSA) and the Committee on Space Research (COSPAR), promote the peaceful use of outer space, coordinate international research efforts, and establish standards for mission planning and data sharing.

36. **Question**: How did India's Chandrayaan-2 mission contribute to lunar exploration?
    Answer: Launched in 2019, Chandrayaan-2 aimed to explore the Moon's south pole. Despite the Vikram lander failing to make a soft landing, the orbiter continues to provide valuable data on lunar topography, mineralogy, and exosphere.

37. **Question**: How will future international collaborations shape the next era of robotic space exploration?
    Answer: Future international collaborations will pool resources, expertise, and technologies from multiple countries to tackle ambitious goals. Projects like the Lunar Gateway, the Mars Sample Return mission, and joint asteroid exploration initiatives will benefit from these partnerships.

38. **Question**: What role did the International Rosetta mission play in studying comets?
    Answer: The Rosetta mission provided detailed data on Comet 67P/Churyumov-Gerasimenko, including its composition, structure, and activity. It advanced our understanding of comets and their role in the solar system's formation. (See Figure 3.4.)

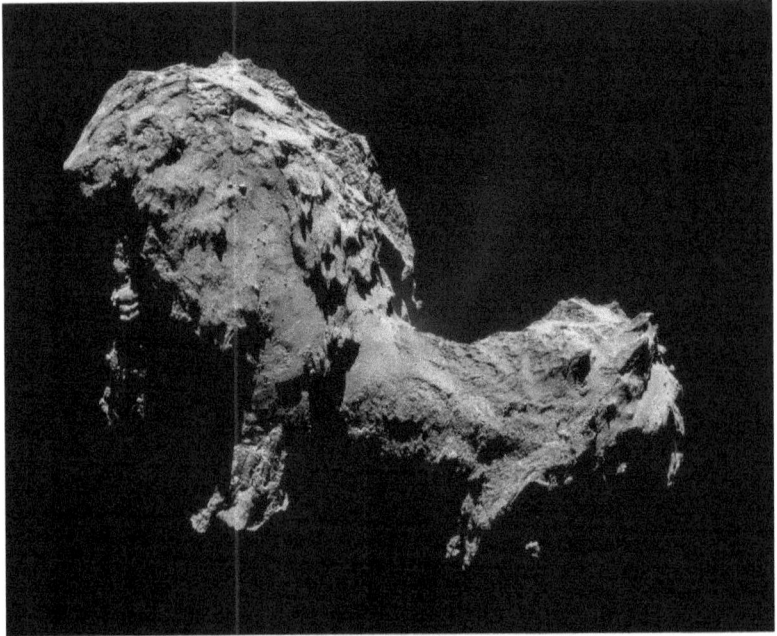

FIGURE 3.4 Comet 67P on September 19, 2014—NavCam mosaic. This mosaic of images taken by the Rosetta spacecraft's navigation camera (NavCam) shows Comet 67P/Churyumov-Gerasimenko. Source: Wikimedia Commons.

39. **Question**: How did the collaboration between NASA and ESA contribute to the success of the Hubble Space Telescope?
    Answer: The collaboration between NASA and ESA allowed for the pooling of resources and expertise, leading to the successful launch and operation of the Hubble Space Telescope. This partnership enabled significant discoveries in astronomy and astrophysics.

40. **Question**: What is the significance of the joint mission between NASA and ISRO for Mars exploration?
 Answer: The joint mission between NASA and ISRO, the Mars Orbiter Mission (Mangalyaan), has provided valuable data on the Martian atmosphere, surface, and climate. This collaboration has enhanced our understanding of Mars and demonstrated the benefits of international cooperation.

41. **Question**: What is the significance of China's Tianwen-1 mission to Mars?
 Answer: Tianwen-1, launched in 2020, is China's first interplanetary mission. It successfully deployed the Zhurong rover on Mars, making China the third country to land on Mars. The mission aims to study the Martian surface, atmosphere, and potential for life.

42. **Question**: How did the Cassini-Huygens mission improve our knowledge of Saturn and its moons?
 Answer: The Cassini-Huygens mission provided detailed observations of Saturn's rings, atmosphere, and magnetic field. The Huygens probe revealed the surface and atmosphere of Titan, Saturn's largest moon, enhancing our knowledge of the Saturnian system.

43. **Question**: What were the key findings of the Dawn mission at Ceres and Vesta?
 Answer: The Dawn mission provided detailed data on the composition and structure of the dwarf planet Ceres and the asteroid Vesta. It revealed evidence of water-ice, hydrated minerals, and complex geological processes, enhancing our understanding of these celestial bodies.

44. **Question**: How have advancements in miniaturization and technology impacted robotic space missions?
 Answer: Advancements in miniaturization and technology have enabled the development of smaller, more efficient robotic spacecraft and instruments. These innovations have reduced costs, increased mission capabilities, and allowed for more frequent and diverse space missions.

45. **Question**: What is the significance of the Mars Sample Return mission?
 Answer: The Mars Sample Return mission aims to collect samples from the Martian surface and return them to Earth for detailed analysis. This mission will provide unprecedented insights into Mars's geology, climate, and potential for past life, significantly advancing our understanding of the Red Planet.

46. **Question**: How have international collaborations contributed to the study of asteroids and comets?
    Answer: International collaborations, such as the Rosetta mission and the Hayabusa missions, have provided detailed data on the composition, structure, and activity of asteroids and comets. These missions have advanced our understanding of these primordial bodies and their role in the solar system's formation.

47. **Question**: What role did the Galileo mission play in studying Jupiter and its moons?
    Answer: The Galileo mission provided detailed observations of Jupiter's atmosphere, magnetic field, and its major moons, including Io, Europa, Ganymede, and Callisto. It revealed evidence of volcanic activity, subsurface oceans, and complex geological processes, enhancing our understanding of the Jovian system.

48. **Question**: How did India's Mangalyaan mission impact Mars exploration?
    Answer: India's Mars Orbiter Mission (Mangalyaan), launched in 2013, made India the first country to successfully reach Mars on its first attempt. The mission has provided valuable data on Martian weather, surface conditions, and atmospheric composition.

49. **Question**: How did the Parker Solar Probe contribute to our knowledge of the sun?
    Answer: The Parker Solar Probe, launched in 2018, is providing unprecedented data on the sun's outer atmosphere and solar wind. Its close approach to the sun is revealing insights into the processes that drive solar activity and influence space weather.

50. **Question**: What is the significance of the James Webb Space Telescope for robotic space exploration?
    Answer: The James Webb Space Telescope, set to launch in 2021, will provide unprecedented observations of the universe in the infrared spectrum. Its advanced instruments and large mirror will enable detailed studies of distant galaxies, star formation, and exoplanets, significantly advancing our understanding of the cosmos. (See Figure 3.5.)

FIGURE 3.5 Its advanced instruments and large mirror of the James Webb Space Telescope enable detailed studies of distant galaxies, star formation, and exoplanets, significantly advancing our understanding of the cosmos. Source: Wikimedia Commons.

51. **Question**: How have international collaborations contributed to the study of exoplanets?

    Answer: International collaborations, such as the Kepler mission and the Transiting Exoplanet Survey Satellite (TESS), have provided detailed data on the characteristics and distribution of exoplanets. These missions have advanced our understanding of planetary systems and the potential for habitable worlds beyond our solar system.

52. **Question**: What role did the Chandra X-ray Observatory play in studying the universe?

    Answer: The Chandra X-ray Observatory has provided detailed observations of high-energy phenomena in the universe, such as black holes, supernova remnants, and galaxy clusters. Its data has enhanced our understanding of the dynamic and energetic processes that shape the cosmos.

53. **Question**: How did the international collaboration on the Event Horizon Telescope contribute to our understanding of black holes?
    Answer: The international collaboration on the Event Horizon Telescope produced the first-ever image of a black hole, providing direct evidence of its existence and characteristics. This achievement has advanced our understanding of black holes and their role in the universe.

54. **Question**: How will future international collaborations shape the next era of robotic space exploration?
    Answer: Future international collaborations will pool resources, expertise, and technologies from multiple countries to tackle ambitious goals. Projects like the Lunar Gateway, the Mars Sample Return mission, and joint asteroid exploration initiatives will benefit from these partnerships, driving the advancement of space exploration.

Robotic explorers have revolutionized our understanding of the solar system and the broader universe. From the pioneering Mars rovers to the groundbreaking Voyager and New Horizons missions and the vital international collaborations, these robotic missions have pushed the boundaries of what we know about space. The achievements of these missions have provided unprecedented insights into the geology, climate, and potential habitability of other planets and moons, transforming our understanding of our place in the cosmos. The legacy of these robotic explorers is evident in the wealth of scientific data they have provided and the technological advancements they have spurred. As technology advances and international partnerships strengthen, the future of robotic space exploration promises even greater discoveries and achievements. The continued success of these missions underscores the importance of robotic explorers in our quest to understand the universe, paving the way for future human exploration and the potential for finding life beyond Earth.

REFERENCES

[Burgess07] Burgess, Colin, and Francis French, *In the Shadow of the Moon: A Challenging Journey to Tranquility, 1965–1969*, University of Nebraska Press, 2007.

[Cain18] Cain, Fraser, *The Universe Today Ultimate Guide to Viewing the Cosmos: Everything You Need to Know to Become an Amateur Astronomer*, Page Street Publishing, 2018.

[Dick06] Dick, Steven J., and Roger D. Launius, eds., *Critical Issues in the History of Spaceflight*, NASA, 2006.

[Neufeld07] Neufeld, Michael J., *Von Braun: Dreamer of Space, Engineer of War*, Alfred A. Knopf, 2007.

[Squyres05] Squyres, Steven, *Roving Mars: Spirit, Opportunity, and the Exploration of the Red Planet*, Hyperion, 2005.

[Wilford90] Wilford, John Noble, *Mars Beckons: The Mysteries, the Challenges, the Expectations of Our Next Great Adventure in Space*, Alfred A. Knopf, 1990.

CHAPTER 4

SPACECRAFT AND TECHNOLOGIES

INTRODUCTION

Spacecraft and their technologies have evolved significantly over the past decades, enabling humanity to explore the far reaches of our solar system and beyond. This chapter delves into the progression of rocket technology, the engineering feats from the Mercury program to the Artemis missions, and the innovations in propulsion and materials that have revolutionized space travel. These key milestones, technological advancements, and international collaborations have driven the development of spacecraft. From the earliest rockets to the sophisticated systems of today, this chapter provides an in-depth look at how spacecraft technology has shaped and will continue to shape our understanding of the universe.

EVOLUTION OF ROCKET TECHNOLOGY

1. **Question**: What were the earliest forms of rocket technology?
   Answer: The earliest forms of rocket technology date back to ancient China, where gunpowder-filled tubes were used as military weapons and fireworks. These simple solid-fuel rockets, known as "fire arrows," were propelled by the explosive force of burning gunpowder and represent the initial steps in the development of rocketry.

2. **Question**: How did World War II influence rocket technology?
   Answer: World War II significantly advanced rocket technology, particularly through the development of the German V-2 rocket. The V-2, designed by Wernher von Braun, was the world's first

long-range guided ballistic missile and marked the transition from basic rocketry to advanced propulsion systems. Its technology laid the groundwork for post-war space exploration efforts.

3. **Question**: What role did the Soviet R-7 rocket play in the Space Race?
    Answer: The Soviet R-7 rocket, developed by Sergei Korolev, was pivotal in the Space Race. It launched Sputnik 1, the first artificial satellite, in 1957 and Yuri Gagarin, the first human in space, in 1961. The R-7's reliable design and powerful thrust capabilities made it a cornerstone of early Soviet space achievements.

4. **Question**: How did the United States develop its rocket technology during the early space age?
    Answer: The United States developed its rocket technology through projects such as Redstone, Atlas, and Titan. Redstone rockets launched the first American astronauts, while Atlas rockets were used for orbital missions. The development of the Saturn V rocket, under the guidance of aerospace engineer Wernher von Braun, culminated in the successful Apollo moon landings.

5. **Question**: What was the significance of the Saturn V rocket?
    Answer: The Saturn V rocket was significant for its role in the Apollo program, enabling humans to reach the moon. It remains the most powerful rocket ever built, with a thrust capacity of 7.5 million pounds. The Saturn V's reliable performance and engineering excellence were critical to the success of the Apollo missions.

6. **Question**: How did the Space Shuttle program revolutionize space travel?
    Answer: The Space Shuttle program, initiated by NASA in the 1980s, revolutionized space travel by introducing reusable spacecraft. The Shuttle's ability to carry large payloads, deploy satellites, and facilitate scientific experiments in orbit marked a significant advancement in space technology. Its reusability aimed to reduce the cost of space missions and pave the way for more frequent launches.

7. **Question**: What advancements in rocket technology were achieved by SpaceX?
    Answer: SpaceX achieved significant advancements in rocket technology with the development of the Falcon 1, Falcon 9, and Falcon Heavy rockets. Key innovations include the reuse of the

first stage of the Falcon 9, reducing launch costs, and the successful deployment of heavy payloads with the Falcon Heavy. SpaceX's advancements have revitalized the commercial space industry.

8. **Question**: What is the significance of the Artemis program's Space Launch System (SLS)?

   Answer: The Artemis program's SLS is significant as NASA's next-generation rocket designed to return humans to the moon and enable deep space exploration. The SLS boasts powerful thrust capabilities and is intended to carry astronauts, cargo, and the Orion spacecraft to lunar orbit and beyond, supporting long-term human presence on the moon and future missions to Mars. (See Figure 4.1.)

FIGURE 4.1 The powerful SLS rocket on the launch pad, ready for the Artemis I mission. This image showcases the most powerful rocket NASA has built, designed to take humans deeper into space than ever before. Source: NASA.

9. **Question**: How have international collaborations influenced the evolution of rocket technology?

   Answer: International collaborations have played a crucial role in the evolution of rocket technology. Programs like the European Ariane rockets, Russia's Soyuz, and joint ventures such as the International Space Station (ISS) have pooled resources and expertise from multiple countries. These collaborations have led to advancements in reliability, cost-efficiency, and technological innovation in rocketry.

10. **Question**: How has China's Long March rocket series contributed to its space program?
    Answer: The Long March rocket series has been instrumental in China's space endeavors, supporting missions from launching satellites to crewed spaceflights. The Long March 5, for example, enabled the Chang'e 5 lunar sample return mission and the Tianwen-1 Mars mission.

11. **Question**: What are some key challenges in modern rocket technology?
    Answer: Key challenges in modern rocket technology include improving fuel efficiency, reducing launch costs, enhancing reusability, and ensuring the safety and reliability of missions. Addressing environmental impacts, such as space debris and emissions from rocket launches, is also a critical concern for the sustainability of space exploration.

SPACECRAFT ENGINEERING: FROM MERCURY TO ARTEMIS

12. **Question**: What were the primary goals of the Mercury program?
    Answer: The primary goals of the Mercury program, NASA's first human spaceflight program, were to orbit a manned spacecraft around Earth, investigate human ability to function in space, and recover both the astronaut and spacecraft safely. The program achieved several milestones, including sending the first American, Alan Shepard, into space and John Glenn into orbit.

13. **Question**: How did the Gemini program advance human space-flight capabilities?
    Answer: The Gemini program advanced human spaceflight capabilities by developing techniques for long-duration missions, spacewalks, and orbital rendezvous and docking. These techniques were essential for the success of the Apollo program. Gemini missions also conducted scientific experiments and tested the physiological effects of space travel on astronauts.

14. **Question**: What advancements did India's Geosynchronous Satellite Launch Vehicle (GSLV) program bring to space technology?
    Answer: The GSLV program enabled India to launch heavier payloads into higher orbits. Key achievements include deploying communication satellites and launching the Chandrayaan-2 mission, demonstrating India's growing capabilities in space technology.

15. **Question**: What engineering challenges did the Apollo program overcome?

    Answer: The Apollo program overcame numerous engineering challenges, including the development of the Saturn V rocket, the Lunar Module for moon landings, and the Command and Service Module for Earth return. Engineers had to ensure the reliability of life support systems, navigation, and reentry technologies, as well as manage the complex logistics of lunar missions.

16. **Question**: How did the Space Shuttle design differ from previous spacecraft?

    Answer: The Space Shuttle design differed from previous spacecraft in its reusability and versatility. Unlike the single-use Mercury, Gemini, and Apollo capsules, the Shuttle featured a reusable orbiter, solid rocket boosters, and an external fuel tank. It was designed to carry both crew and cargo, deploy satellites, and serve as a laboratory for scientific research in space.

17. **Question**: What role did the International Space Station (ISS) play in advancing spacecraft engineering?

    Answer: The ISS played a critical role in advancing spacecraft engineering by serving as a platform for long-term human habitation and scientific research in microgravity. Its construction involved complex engineering feats, including docking and assembly of modules in orbit. The ISS has provided valuable data on the effects of prolonged spaceflight on humans and tested technologies for future deep space missions.

18. **Question**: What are the unique features of China's Tiangong space station?

    Answer: China's Tiangong space station, launched in 2022, features modular design, advanced life support systems, and capabilities for long-term human habitation. It represents China's ambition to establish a continuous human presence in low Earth orbit.

19. **Question**: How has the Orion spacecraft contributed to NASA's deep space exploration goals?

    Answer: The Orion spacecraft, developed for NASA's Artemis program, is designed for deep space exploration, including missions to the moon and Mars. Orion features advanced life support, navigation, and propulsion systems, and can carry astronauts on long-duration missions. Its capabilities support the goal of establishing a sustainable human presence on the moon and preparing for future Mars exploration. (See Figure 4.2.)

FIGURE 4.2 Technicians assembling the Orion spacecraft, which is set to carry astronauts to the Moon and beyond as part of the Artemis missions. This spacecraft is crucial for deep space exploration. Source: NASA.

20. **Question**: What engineering innovations have been introduced in the Artemis program?

    Answer: The Artemis program has introduced several engineering innovations, including the Space Launch System (SLS) for powerful deep space missions, the Orion spacecraft for crewed exploration, and the Gateway lunar orbiting outpost. These innovations aim to enhance mission flexibility, safety, and sustainability, supporting long-term lunar exploration and eventual human missions to Mars.

21. **Question**: How has SpaceX's Dragon spacecraft impacted commercial spaceflight?

    Answer: SpaceX's Dragon spacecraft has significantly impacted commercial spaceflight by providing reliable cargo and crew transport to the ISS. The Dragon's reusability and cost-effectiveness have lowered the barriers to space access for commercial and government customers. Its success has paved the way for future private space missions and the development of the Starship spacecraft for deep space travel.

22. **Question**: What role did the Hubble Space Telescope play in spacecraft engineering?

    Answer: The Hubble Space Telescope played a pivotal role in spacecraft engineering by demonstrating the feasibility of on-orbit servicing and repair. Astronauts conducted multiple servicing missions to upgrade Hubble's instruments and extend its operational life. These missions showcased the importance of maintaining and improving space-based observatories and influenced the design of future telescopes.

23. **Question**: How did the James Webb Space Telescope advance spacecraft engineering?

    Answer: The James Webb Space Telescope advanced spacecraft engineering through its innovative design and construction. It features a large, segmented primary mirror and a sunshield to protect its instruments from solar heat. The telescope's deployment and operation in a distant orbit require precise engineering and advanced technologies, setting new standards for space-based observatories.

24. **Question**: How did India's satellite navigation system (NavIC) enhance technological capabilities?

    Answer: The NavIC provides accurate position information services to users in India and the surrounding region. It enhances navigation, disaster management, and scientific research capabilities, showcasing India's advancements in satellite technology.

INNOVATIONS IN PROPULSION AND MATERIALS

25. **Question**: What are the primary types of propulsion systems used in space exploration?

    Answer: The primary types of propulsion systems used in space exploration include chemical propulsion (solid and liquid rockets), electric propulsion (ion and Hall effect thrusters), and nuclear propulsion (thermal and electric). Each type offers different advantages in terms of thrust, efficiency, and suitability for various mission profiles.

26. **Question**: How has ion propulsion been used in space missions?

    Answer: Ion propulsion has been used in space missions for its high efficiency and ability to provide continuous low-thrust acceleration over long durations. Notable missions include NASA's Deep Space 1, which demonstrated ion propulsion, and the Dawn mission, which used ion thrusters to explore the asteroid belt and study Vesta and Ceres.

27. **Question**: What are the advantages of using electric propulsion for space travel?
Answer: Electric propulsion offers several advantages for space travel, including higher efficiency, longer operational lifetimes, and the ability to achieve higher velocities over extended periods. This makes it ideal for deep space missions where efficient use of propellant is critical. Electric propulsion systems are also more compact, allowing for larger payloads.

28. **Question**: How has the development of reusable rockets impacted space exploration?
Answer: The development of reusable rockets has significantly impacted space exploration by reducing launch costs and increasing the frequency of missions. Companies like SpaceX and Blue Origin have pioneered reusable rocket technology, demonstrating the ability to land and relaunch first-stage boosters. This innovation has made space access more economical and sustainable.

29. **Question**: What is the significance of the development of composite materials in spacecraft construction?
Answer: The development of composite materials in spacecraft construction is significant for their high strength-to-weight ratio, durability, and resistance to extreme temperatures. Composites are used in various spacecraft components, including structural frames, heat shields, and fuel tanks, enhancing performance and reducing overall weight.

30. **Question**: How have advancements in thermal protection systems improved spacecraft safety?
Answer: Advancements in thermal protection systems have improved spacecraft safety by enabling them to withstand the intense heat and friction of atmospheric reentry. Modern materials, such as reinforced carbon-carbon and ablative heat shields, protect spacecraft from burning up upon reentry, ensuring the safe return of astronauts and equipment.

31. **Question**: What role does 3D printing play in spacecraft engineering?
Answer: 3D printing plays an increasingly important role in spacecraft engineering by allowing the rapid and cost-effective production of complex components. It enables the creation of lightweight, customized parts and reduces the need for extensive supply chains.

3D printing is used for manufacturing rocket engines, structural components, and even habitat modules for space exploration.

32. **Question**: How has the development of advanced propulsion technologies like nuclear thermal propulsion impacted space travel?

    Answer: The development of advanced propulsion technologies like nuclear thermal propulsion has the potential to significantly impact space travel by providing higher thrust and efficiency compared to chemical rockets. Nuclear thermal propulsion uses nuclear reactions to heat propellant, offering greater performance for deep space missions and reducing travel times to distant destinations like Mars.

33. **Question**: What are the benefits of using solar sails for space exploration?

    Answer: Solar sails offer benefits for space exploration by harnessing the pressure of sunlight for propulsion. This method provides continuous acceleration without the need for traditional fuel, making it ideal for long-duration missions. Solar sails can enable exploration of distant regions of the solar system and interstellar space with minimal propellant requirements. (See Figure 4.3.)

FIGURE 4.3 The LightSail 2 mission, developed by The Planetary Society, demonstrates the use of solar sails for space propulsion. Solar sails offer continuous acceleration by harnessing the pressure of sunlight, ideal for long-duration missions exploring distant regions of the solar system and interstellar space. Source: Wikimedia Commons.

34. **Question**: How have advancements in fuel technology improved rocket performance?
    Answer: Advancements in fuel technology have improved rocket performance by increasing efficiency, thrust, and safety. The development of high-energy propellants, cryogenic fuels, and green propellants has enhanced the capabilities of rockets, enabling them to carry heavier payloads, achieve higher velocities, and reduce environmental impact.

35. **Question**: What innovations have been made in spacecraft navigation systems?
    Answer: Innovations in spacecraft navigation systems include the development of autonomous navigation, precision timing, and deep space communication technologies. These advancements allow spacecraft to navigate accurately without relying solely on Earth-based tracking, improving mission efficiency and enabling more complex maneuvers.

36. **Question**: How has artificial intelligence (AI) been integrated into spacecraft systems?
    Answer: AI has been integrated into spacecraft systems to enhance autonomy, decision-making, and data analysis. AI algorithms are used for tasks such as anomaly detection, path planning, and scientific data processing, allowing spacecraft to operate more independently and efficiently in the harsh environment of space.

37. **Question**: What is the role of cryogenic technology in rocket propulsion?
    Answer: Cryogenic technology plays a crucial role in rocket propulsion by enabling the use of supercooled liquid propellants like liquid hydrogen and liquid oxygen. These propellants offer high energy density and efficiency, making them ideal for powerful launch vehicles like the Saturn V and the Space Launch System (SLS).

38. **Question**: How have advancements in miniaturization impacted spacecraft design?
    Answer: Advancements in miniaturization have impacted spacecraft design by allowing the development of smaller, lighter, and more efficient components. This has led to the rise of CubeSats and small satellites, which can perform scientific and commercial missions at a fraction of the cost of traditional spacecraft.

39. **Question**: What is the significance of the development of hybrid rockets?
    Answer: The development of hybrid rockets, which use a combination of solid and liquid propellants, is significant for their simplicity, safety, and controllability. Hybrid rockets offer a balance between the high thrust of solid rockets and the controllable burn of liquid rockets, making them suitable for various applications, including space tourism and suborbital flights.

40. **Question**: How has the use of advanced materials like graphene impacted spacecraft engineering?
    Answer: The use of advanced materials like graphene has impacted spacecraft engineering by offering exceptional strength, flexibility, and thermal conductivity. Graphene's properties make it ideal for applications such as lightweight structural components, thermal management systems, and advanced electronics, enhancing the overall performance and durability of spacecraft.

41. **Question**: What role do ion thrusters play in space missions?
    Answer: Ion thrusters play a crucial role in space missions by providing highly efficient, low-thrust propulsion for long-duration missions. They are used for tasks such as station-keeping, orbital adjustments, and deep space exploration. Ion thrusters have been successfully employed in missions like NASA's Dawn and ESA's BepiColombo. (See Figure 4.4.)

FIGURE 4.4 This image shows NASA's NEXT (NASA's Evolutionary Xenon Thruster) ion propulsion system being tested. Ion thrusters provide highly efficient, low-thrust propulsion for long-duration space missions, used in tasks such as station-keeping, orbital adjustments, and deep space exploration. Source: Wikimedia Commons.

42. **Question**: How has the development of autonomous spacecraft impacted space exploration?
    Answer: The development of autonomous spacecraft has impacted space exploration by enabling more complex and distant missions. Autonomous systems can perform tasks without real-time human intervention, making decisions based on sensor data and predefined algorithms. This capability is essential for missions to distant planets, moons, and asteroids.

43. **Question**: What are the benefits of using solid-state batteries in spacecraft?
    Answer: Solid-state batteries offer benefits for spacecraft by providing higher energy density, improved safety, and longer lifespans compared to traditional liquid electrolyte batteries. Their compact size and robustness make them ideal for space applications, where reliability and efficiency are critical.

44. **Question**: How have advancements in thermal management systems improved spacecraft performance?
    Answer: Advancements in thermal management systems have improved spacecraft performance by ensuring optimal temperature control for instruments and components. Techniques such as radiators, heat pipes, and phase-change materials help dissipate excess heat and maintain stable operating conditions, enhancing the longevity and reliability of spacecraft.

45. **Question**: What is the significance of the development of plasma propulsion?
    Answer: The development of plasma propulsion is significant for its potential to provide efficient, high-thrust propulsion for deep space missions. Plasma thrusters, such as Hall effect thrusters, use ionized gas to generate thrust, offering advantages in specific impulse and fuel efficiency over traditional chemical propulsion.

46. **Question**: How has the integration of advanced sensors improved spacecraft capabilities?
    Answer: The integration of advanced sensors has improved spacecraft capabilities by providing precise measurements and real-time data for navigation, scientific research, and system monitoring. Sensors such as LIDAR, spectrometers, and accelerometers enhance situational awareness and enable more accurate and efficient mission operations.

47. **Question**: What role does robotics play in spacecraft maintenance and repair?
    Answer: Robotics plays a crucial role in spacecraft maintenance and repair by enabling remote operations and servicing of space assets. Robotic arms, such as those used on the ISS, can perform tasks like satellite repair, module assembly, and scientific instrument deployment, reducing the need for human spacewalks and extending the operational life of spacecraft.

48. **Question**: How have advancements in communications technology impacted spacecraft operations?
    Answer: Advancements in communications technology have impacted spacecraft operations by enabling faster, more reliable data transmission between spacecraft and Earth. Innovations such as laser communication systems provide higher bandwidth and lower latency, supporting complex missions and real-time data analysis.

49. **Question**: What is the significance of the development of nuclear electric propulsion?
    Answer: The development of nuclear electric propulsion is significant for its potential to provide long-duration, high-efficiency propulsion for deep space missions. By using nuclear reactors to generate electricity for ion or plasma thrusters, this technology offers the capability to explore distant destinations like Mars and beyond with greater speed and efficiency.

50. **Question**: How has the use of advanced alloys improved spacecraft durability?
    Answer: The use of advanced alloys has improved spacecraft durability by providing materials that can withstand the extreme conditions of space, such as high radiation, temperature fluctuations, and mechanical stress. These alloys enhance the structural integrity and longevity of spacecraft, making them more reliable for long-term missions.

51. **Question**: What role does artificial intelligence play in mission planning and execution?
    Answer: Artificial intelligence plays a critical role in mission planning and execution by optimizing trajectories, managing resources, and analyzing scientific data. AI algorithms can simulate various mission scenarios, identify potential risks, and make autonomous decisions, improving the efficiency and success rates of space missions.

52. **Question**: How has the development of lightweight materials impacted spacecraft design?

    Answer: The development of lightweight materials has impacted spacecraft design by reducing launch costs and increasing payload capacity. Materials such as carbon fiber composites and aerogels offer high strength-to-weight ratios, allowing for more efficient spacecraft structures and enabling more ambitious mission designs.

53. **Question**: What are the benefits of using modular spacecraft architectures?

    Answer: Modular spacecraft architectures offer benefits such as flexibility, scalability, and cost-efficiency. By designing spacecraft with interchangeable modules, missions can be tailored to specific objectives, upgraded with new technologies, and repaired or expanded in orbit, enhancing overall mission versatility and longevity. (See Figure 4.5.)

FIGURE 4.5 The image illustrates the modular buildup of the Lunar Orbital Platform-Gateway (LOP-G) by Sierra Nevada Corporation. Modular spacecraft architectures offer flexibility, scalability, and cost-efficiency, allowing for tailored missions, technological upgrades, and in-orbit repairs or expansions. Source: Wikimedia Commons.

54. **Question**: How has the development of propulsion systems for small satellites impacted space exploration?

    Answer: The development of propulsion systems for small satellites has impacted space exploration by enabling precise maneuvers, station-keeping, and extended mission lifetimes for small spacecraft. These systems, such as miniaturized ion thrusters and cold gas thrusters, provide small satellites with greater autonomy and functionality, supporting a wide range of scientific, commercial, and defense applications.

The evolution of rocket technology, spacecraft engineering, and propulsion innovations has been instrumental in advancing space exploration. From the pioneering efforts of early rocketry to the sophisticated missions of today, these technological advancements have enabled humanity to explore and understand the cosmos in unprecedented ways. The collaboration between international space agencies and private companies has accelerated the pace of discovery, paving the way for future missions that promise to unlock even more secrets of the universe.

As we continue to push the boundaries of space exploration, the technologies developed today will serve as the foundation for the next generation of explorers, ensuring that our quest for knowledge and discovery continues. The advancements in rocket technology and spacecraft engineering not only enhance our ability to conduct scientific research but also inspire future generations to dream big and pursue careers in STEM fields. With ongoing innovations in propulsion systems and materials, we are poised to explore deeper into space, potentially even reaching other star systems. The future of space exploration is bright, and the lessons learned from our past and present endeavors will guide us on this exciting journey.

REFERENCES

[Chaikin94] Chaikin, Andrew, *A Man on the Moon: The Voyages of the Apollo Astronauts*, Penguin Books, 1994.

[Harland05a] Harland, David M., *Space Systems Failures: Disasters and Rescues of Satellites, Rockets and Space Probes*, Springer-Praxis, 2005.

[Jenkins07] Jenkins, Dennis R., *Space Shuttle: The History of the National Space Transportation System: The First 100 Missions*, Specialty Press, 2007.

[Neufeld07] Neufeld, Michael J., *Von Braun: Dreamer of Space, Engineer of War*, Alfred A. Knopf, 2007.

[Ward15] Ward, Jonathan, *Rocket Ranch: The Nuts and Bolts of the Apollo Moon Program at Kennedy Space Center*, Springer-Praxis, 2015.

[Wilford90] Wilford, John Noble, *Mars Beckons: The Mysteries, the Challenges, the Expectations of Our Next Great Adventure in Space*, Alfred A. Knopf, 1990.

[Winter05] Winter, Frank H., and William P. Barry, *Rockets and People: Volume I*, NASA History Series SP-2005-4110, NASA, 2005.

CHAPTER 5

THE INTERNATIONAL SPACE STATION AND BEYOND

INTRODUCTION

The International Space Station (ISS) represents one of humanity's most ambitious and successful endeavors in space exploration. Serving as a hub for scientific research, international cooperation, and technological innovation, the ISS embodies the collaborative spirit and technical prowess necessary to maintain a permanent human presence in low Earth orbit. This chapter delves into the birth and operation of the ISS, highlighting the pivotal roles of the United States, Russia, Europe, Japan, and Canada. Explore the intricate process of designing, constructing, and maintaining the ISS, as well as the significant scientific advancements and educational initiatives it has supported. Additionally, this chapter will discuss the future plans for lunar bases and Mars missions, illustrating how the lessons learned from the ISS are shaping the next era of space exploration.

The ISS stands as a testament to what can be achieved through international collaboration, with contributions from various space agencies combining to create a state-of-the-art research facility in space. From its early concepts to the continuous upgrades that keep it operational, the ISS is a dynamic platform that adapts to new scientific challenges and technological opportunities. As we look to the future, the partnerships and innovations fostered on the ISS will pave the way for ambitious projects such as lunar bases and human missions to Mars.

THE BIRTH AND OPERATION OF THE ISS

1. **Question**: What were the initial concepts that led to the creation of the ISS?

   Answer: The initial concepts that led to the creation of the ISS originated in the early 1980s when NASA proposed the Space Station Freedom project. This idea evolved through international collaboration and a desire to create a permanent human presence in low Earth orbit for scientific research, technology development, and fostering international partnerships. The ISS was formally initiated through agreements between NASA, Roscosmos, ESA, JAXA, and CSA. (See Figure 5.1.)

FIGURE 5.1 This image shows the ISS orbiting Earth, highlighting its massive structure and the intricate components that make up the space station. This photograph was taken during a space shuttle mission. Source: NASA.

2. **Question**: How did international collaboration shape the design and construction of the ISS?

   Answer: International collaboration was crucial in shaping the design and construction of the ISS. Each partner country contributed modules, technology, and expertise. The United States provided key elements like the Unity and Destiny modules, Russia contributed the Zarya and Zvezda modules, Europe developed the Columbus laboratory, Japan built the Kibo module, and Canada provided the Canadarm2 robotic system. This collaborative effort ensured a diverse and capable space station.

3. **Question**: What were the key milestones in the assembly of the ISS?
    Answer: Key milestones in the assembly of the ISS include the launch of the Russian Zarya module in 1998, which marked the start of the station's construction. Subsequent milestones included the addition of the Unity module, the first US component, the Zvezda service module, and the installation of the solar arrays. Major laboratories like Destiny, Columbus, and Kibo were added over the years, and the assembly was completed with the installation of critical components like the Integrated Truss Structure and the Node 3 (Tranquility) module.

4. **Question**: How is the ISS maintained and operated?
    Answer: The ISS is maintained and operated through continuous collaboration among international space agencies. Mission control centers in the United States, Russia, Europe, Japan, and Canada coordinate activities. Crews conduct regular maintenance, scientific experiments, and technological demonstrations. Cargo resupply missions from vehicles like SpaceX's Dragon, Northrop Grumman's Cygnus, and Russia's Progress provide necessary supplies, equipment, and scientific instruments.

5. **Question**: What role do astronauts play in the operation of the ISS?
    Answer: Astronauts play a vital role in the operation of the ISS. They conduct scientific experiments, maintain the station's systems, and perform spacewalks to install new equipment and repair existing components. Astronauts also engage in international collaboration by working alongside crewmates from different countries, fostering a spirit of cooperation and shared goals.

6. **Question**: How has the ISS contributed to scientific research?
    Answer: The ISS has significantly contributed to scientific research by providing a unique microgravity environment for experiments in fields such as biology, physics, astronomy, and materials science. Research conducted on the ISS has led to advancements in medical treatments, improved materials, and a better understanding of fundamental scientific principles. Studies on the ISS have also provided insights into the long-term effects of space travel on the human body.

7. **Question**: What technological advancements have been tested on the ISS?
    Answer: The ISS has served as a testbed for numerous technological advancements, including life support systems, advanced robotics, and new materials. Technologies like the Water Recovery

System, which recycles wastewater into drinkable water, and the Advanced Closed Loop System for air revitalization have been crucial for long-duration space missions. Additionally, the testing of autonomous robotic systems like the Robonaut and the Astrobee robots has paved the way for future space exploration.

8. **Question**: How has the ISS supported educational initiatives?
   Answer: The ISS has supported educational initiatives through programs that engage students and educators worldwide. Initiatives like the "Year of Education on Station" and the "ISS National Lab" provide opportunities for students to design and conduct experiments in space. Astronauts also participate in live broadcasts and educational events, inspiring the next generation of scientists, engineers, and explorers.

9. **Question**: What is the significance of the ISS in terms of international cooperation?
   Answer: The ISS is a symbol of international cooperation, demonstrating how countries can work together to achieve common goals in space exploration. The collaboration between NASA, Roscosmos, ESA, JAXA, and CSA, along with contributions from other countries, has fostered diplomatic relationships and shared scientific and technological advancements. The ISS serves as a model for future international partnerships in space exploration.

10. **Question**: How has the ISS evolved over time?
    Answer: The ISS has evolved through continuous upgrades and additions of new modules and technologies. Initially focused on assembly, the station has transitioned to a fully operational research laboratory. Upgrades like the addition of new solar arrays, improved life support systems, and advanced scientific instruments have enhanced its capabilities. The ISS continues to evolve to support new research and technological demonstrations, preparing for future exploration missions.

PARTNERSHIPS AND CONTRIBUTIONS OF PARTICIPATING NATIONS

11. **Question**: What contributions has NASA made to the ISS?
    Answer: NASA has made significant contributions to the ISS, including the development and launch of key modules like Unity, Destiny, Harmony, and Tranquility. NASA has also provided

essential infrastructure such as the Integrated Truss Structure, solar arrays, and external stowage platforms. Additionally, NASA oversees many of the station's operations and scientific research programs, coordinating with international partners to ensure the ISS's success.

12. **Question**: How has Roscosmos contributed to the ISS?
    Answer: Roscosmos has been instrumental in the ISS's development, contributing the Zarya and Zvezda modules, which provide essential power, propulsion, and life support systems. Russian Soyuz spacecraft are used for crew transport, and Progress vehicles supply cargo. Roscosmos also plays a key role in maintaining the station's orbit and conducting scientific experiments in Russian modules.

13. **Question**: What role has the European Space Agency (ESA) played in the ISS?
    Answer: The ESA has played a crucial role in the ISS by developing and providing the Columbus laboratory module, which hosts a variety of scientific experiments. ESA also contributed the Automated Transfer Vehicle (ATV) for cargo resupply missions and the European Robotic Arm (ERA) for maintenance and assembly tasks. ESA's participation has enhanced the ISS's scientific capabilities and international collaboration.

14. **Question**: How has the Japan Aerospace Exploration Agency (JAXA) contributed to the ISS?
    Answer: The JAXA has contributed the Kibo laboratory module, which includes a pressurized module, an exposed facility for experiments in space, and a logistics module for storage. JAXA's HTV (H-II Transfer Vehicle) has been vital for delivering supplies, equipment, and scientific instruments to the ISS. JAXA's contributions have expanded the station's research capabilities and international partnerships.

15. **Question**: What contributions has the Canadian Space Agency (CSA) made to the ISS?
    Answer: The CSA has made significant contributions with the development of the Canadarm2, a sophisticated robotic arm used for assembly, maintenance, and docking operations. CSA also provided the Dextre robotic system for delicate repairs and the Mobile Base System for moving equipment along the station's truss. These contributions have been critical for the station's construction and ongoing operations. (See Figure 5.2.)

FIGURE 5.2 The Canadarm2 and Dextre, developed by the Canadian Space Agency (CSA), are critical robotic systems used for assembly, maintenance, and docking operations on the International Space Station (ISS). Source: Wikimedia.

16. **Question**: How have private companies contributed to the ISS?
    Answer: Private companies have contributed to the ISS through cargo resupply missions, crew transport, and the development of new technologies. Companies like SpaceX, Northrop Grumman, and Boeing have provided cargo spacecraft like Dragon, Cygnus, and CST-100 Starliner. SpaceX's Crew Dragon has also transported astronauts to the ISS, marking a new era of commercial spaceflight and expanding access to space.

17. **Question**: What is the significance of international partnerships in ISS research?

    Answer: International partnerships in ISS research have broadened the scope and impact of scientific investigations. Collaborative research projects involving multiple countries have addressed diverse topics such as climate change, medical advancements, and fundamental physics. These partnerships leverage the unique capabilities of the ISS, fostering innovation and shared scientific progress.

18. **Question**: How have joint missions and experiments on the ISS benefited global science?

    Answer: Joint missions and experiments on the ISS have benefited global science by providing a unique platform for research in microgravity. Studies on the ISS have led to advancements in areas such as drug development, materials science, and environmental monitoring. The collaborative nature of ISS research has facilitated knowledge sharing and the development of new technologies that benefit people on Earth.

19. **Question**: What is the role of international funding in the operation of the ISS?

    Answer: International funding plays a crucial role in the operation of the ISS. Partner countries contribute financially to the development, maintenance, and operation of the station. This shared investment ensures the sustainability and success of the ISS, allowing it to serve as a hub for scientific research and international cooperation.

20. **Question**: How have international crews contributed to the success of the ISS?

    Answer: International crews have been essential to the success of the ISS, bringing diverse skills, perspectives, and expertise to the station. Crewmembers from various countries work together to conduct experiments, perform maintenance, and ensure the smooth operation of the station. This collaboration fosters a spirit of unity and shared purpose, enhancing the ISS's mission.

21. **Question**: What impact has the ISS had on international diplomacy?

    Answer: The ISS has had a positive impact on international diplomacy by serving as a symbol of peaceful cooperation and shared achievement. The collaboration between countries with diverse political and cultural backgrounds demonstrates the potential for unity and partnership in pursuit of common goals. The ISS has strengthened diplomatic relations and provided a model for future international projects.

22. **Question**: How has China contributed to the development of its own space station?

    Answer: China has developed the Tiangong space station, which will serve as a hub for scientific research and international collaboration. The Tiangong program includes the successful launch and operation of Tiangong-1 and Tiangong-2 space labs, paving the way for a permanent Chinese space station.

23. **Question**: How have international training programs supported ISS missions?

    Answer: International training programs have supported ISS missions by preparing astronauts and cosmonauts for the challenges of spaceflight. Training programs at facilities like NASA's Johnson Space Center, Russia's Star City, and Europe's EAC (European Astronaut Centre) provide comprehensive training in areas such as spacecraft operations, scientific research, and emergency procedures. These programs ensure that all crew members are well-prepared for their missions.

24. **Question**: How did India's Astrosat mission contribute to space science?

    Answer: Launched in 2015, Astrosat is India's first dedicated multiwavelength space observatory. It studies celestial sources in different wavelengths, including X-rays, UV, and optical. The mission enhances our understanding of high-energy processes in the universe. (See Figure 5.3.)

FIGURE 5.3 Astrosat-1, launched in 2015, is India's first dedicated multiwavelength space observatory. Source: Wikimedia Commons.

25. **Question**: What is the role of the International Space Station Program (ISSP) in coordinating ISS activities?
    Answer: The ISSP coordinates activities related to the development, operation, and utilization of the ISS. The ISSP facilitates collaboration among partner agencies, manages resource allocation, and oversees the planning and execution of missions. The program ensures that the ISS operates smoothly and meets its scientific and operational objectives.

26. **Question**: What role did India play in international space collaborations on the ISS?
    Answer: While India does not have modules on the ISS, it collaborates internationally through satellite launches and scientific experiments. India's space agency, ISRO, has partnered with NASA and other space agencies on various missions, including remote sensing and satellite technology development.

27. **Question**: How have cultural exchanges on the ISS enhanced international understanding?
    Answer: Cultural exchanges on the ISS have enhanced international understanding by fostering communication and collaboration among crewmembers from different countries. Shared experiences and cultural traditions create a sense of camaraderie and mutual respect. These exchanges promote a greater appreciation of diverse perspectives and contribute to the overall success of the ISS mission.

28. **Question**: What challenges have been overcome through international collaboration on the ISS?
    Answer: International collaboration on the ISS has overcome challenges such as technical difficulties, logistical coordination, and cultural differences. Partner agencies have worked together to address issues like module integration, system failures, and emergency situations. This collaboration has demonstrated the resilience and adaptability of the international space community.

29. **Question**: What are the key objectives of China's Tiangong space station program?
    Answer: The Tiangong space station aims to support long-term human presence in space, conduct cutting-edge scientific research, and facilitate international cooperation. It includes modules for living quarters, laboratories, and docking ports for additional spacecraft.

PLANS FOR LUNAR BASES AND MARS MISSIONS

30. **Question**: What are the primary goals of the Artemis program?
    Answer: The primary goals of the Artemis program are to return humans to the moon, establish a sustainable presence, and prepare for future missions to Mars. The program aims to land the first woman and the next man on the lunar surface, conduct scientific research, and develop technologies for long-duration space exploration. Artemis will also build the Lunar Gateway, a space station in lunar orbit to support missions to the moon and beyond.

31. **Question**: How will the Lunar Gateway support future lunar and Mars missions?
    Answer: The Lunar Gateway will support future lunar and Mars missions by serving as a staging point for crewed and uncrewed missions to the lunar surface. It will provide a platform for scientific research, technology demonstrations, and international collaboration. The Gateway's modular design allows for the addition of new capabilities and infrastructure, making it a critical component of long-term exploration plans.

32. **Question**: What role will international partnerships play in the Artemis program?
    Answer: International partnerships will play a crucial role in the Artemis program by contributing technology, expertise, and resources. Countries like Canada, Japan, and European nations are collaborating with NASA to develop components for the Lunar Gateway, lunar landers, and scientific instruments. These partnerships enhance the program's capabilities and ensure a diverse and collaborative approach to lunar exploration.

33. **Question**: How will the Artemis Base Camp on the lunar surface contribute to sustainable exploration?
    Answer: The Artemis Base Camp on the lunar surface will contribute to sustainable exploration by providing a long-term habitat for astronauts, enabling extended missions and continuous scientific research. The base camp will utilize in-situ resource utilization (ISRU) technologies to extract water and other resources from the lunar environment, reducing the need for supplies from Earth. This approach supports sustainable living and prepares for future missions to Mars. (See Figure 5.4.)

FIGURE 5.4 The Artemis Base Camp on the lunar surface will contribute to sustainable exploration by providing a long-term habitat for astronauts, enabling extended missions and continuous scientific research. Source: Wikimedia Commons.

34. **Question**: What technologies are being developed for lunar and Mars habitats?

    Answer: Technologies being developed for lunar and Mars habitats include inflatable modules, advanced life support systems, and radiation protection. Habitat designs focus on sustainability, incorporating systems for recycling air and water, growing food, and managing waste. These technologies are critical for ensuring the health and safety of astronauts on long-duration missions.

35. **Question**: How will robotic missions support human exploration of the moon and Mars?

    Answer: Robotic missions will support human exploration of the moon and Mars by conducting reconnaissance, preparing landing sites, and performing scientific research. Robots can operate in harsh environments and carry out tasks such as drilling, sample collection, and habitat construction. These missions will gather valuable data and pave the way for human explorers.

36. **Question**: What is the significance of NASA's Artemis Accords?

    Answer: The Artemis Accords are a set of principles and guidelines for international cooperation in lunar exploration. They emphasize transparency, interoperability, and the peaceful use of space. The Accords aim to create a framework for collaborative efforts, ensuring that lunar exploration benefits all humanity and adheres to established norms and laws.

37. **Question**: How will the use of in-situ resource utilization (ISRU) impact lunar and Mars missions?
    Answer: The use of ISRU will impact lunar and Mars missions by reducing the dependence on Earth-supplied resources. ISRU technologies can extract water, oxygen, and fuel from the local environment, enabling longer missions and reducing launch costs. This approach supports sustainable exploration and the establishment of permanent bases.

38. **Question**: What role will 3D printing play in the construction of lunar and Mars habitats?
    Answer: 3D printing will play a significant role in the construction of lunar and Mars habitats by enabling the creation of structures using local materials. This technology can produce habitat components, tools, and spare parts on-site, reducing the need for shipments from Earth. 3D printing enhances flexibility and adaptability in habitat construction, supporting long-term exploration goals.

39. **Question**: How will international collaboration shape the future of Mars exploration?
    Answer: International collaboration will shape the future of Mars exploration by pooling resources, expertise, and technology from multiple countries. Collaborative efforts will enhance mission capabilities, share risks and costs, and promote scientific discoveries. Partnerships like the Mars Sample Return mission, involving NASA and ESA, exemplify the benefits of international cooperation.

40. **Question**: What are the primary challenges of human missions to Mars?
    Answer: The primary challenges of human missions to Mars include ensuring crew health and safety, developing reliable life support systems, managing radiation exposure, and providing sufficient food and water. Other challenges include the long duration of the mission, communication delays, and the need for autonomous operation. Addressing these challenges requires advanced technologies and careful planning.

41. **Question**: How will Mars missions benefit from the experience gained on the ISS and lunar missions?
    Answer: Mars missions will benefit from the experience gained on the ISS and lunar missions by leveraging the knowledge and technologies developed in these programs. Lessons learned in areas such as life support, habitat design, and long-duration spaceflight

will inform Mars mission planning. The ISS and lunar missions provide a foundation for testing and refining systems needed for Mars exploration.

42. **Question**: What is the role of space agencies like NASA, ESA, and Roscosmos in Mars exploration?
    Answer: Space agencies like NASA, ESA, and Roscosmos play a leading role in Mars exploration by conducting scientific missions, developing technologies, and fostering international collaboration. Each agency brings unique capabilities and expertise to Mars exploration efforts, working together to achieve common goals. Their coordinated efforts are essential for the success of future Mars missions.

43. **Question**: How will advancements in propulsion technology impact future Mars missions?
    Answer: Advancements in propulsion technology will impact future Mars missions by reducing travel time, increasing efficiency, and enabling more flexible mission profiles. Technologies like nuclear thermal propulsion and ion propulsion offer higher specific impulse and thrust, making them ideal for long-duration missions. Improved propulsion systems will enhance the feasibility and safety of human missions to Mars.

44. **Question**: What scientific objectives are prioritized in Mars exploration missions?
    Answer: Scientific objectives in Mars exploration missions include the search for signs of past or present life, understanding the planet's geology and climate, and studying its potential for human habitation. Missions aim to analyze soil and rock samples, investigate subsurface water, and monitor atmospheric conditions. These objectives provide insights into Mars' history and its suitability for future exploration.

45. **Question**: How will the Mars Sample Return mission contribute to our understanding of Mars?
    Answer: The Mars Sample Return mission will contribute to our understanding of Mars by collecting and returning samples of Martian soil and rock to Earth for detailed analysis. This mission will provide unprecedented insights into the planet's composition, geology, and potential for past life. The returned samples will be studied using advanced laboratory techniques, enhancing our knowledge of Mars and informing future missions.

46. **Question**: What role will autonomous systems play in future Mars missions?
    Answer: Autonomous systems will play a crucial role in future Mars missions by performing tasks without real-time human intervention. These systems can conduct scientific research, operate equipment, and manage resources, allowing missions to function effectively despite communication delays. Autonomous systems enhance mission flexibility and increase the chances of success.

47. **Question**: How will international agreements and policies support lunar and Mars exploration?
    Answer: International agreements and policies will support lunar and Mars exploration by establishing guidelines for cooperation, resource use, and conflict resolution. Agreements like the Artemis Accords promote transparency, safety, and the peaceful use of space. These policies ensure that exploration efforts are conducted responsibly and benefit all humanity.

48. **Question**: What are the potential economic benefits of lunar and Mars exploration?
    Answer: The potential economic benefits of lunar and Mars exploration include the development of new technologies, job creation, and the discovery of valuable resources. Exploration efforts can drive innovation in industries such as robotics, materials science, and energy. Additionally, the extraction of resources like water and minerals could support future space missions and economic activities.

49. **Question**: How will advancements in life support systems impact long-duration space missions?
    Answer: Advancements in life support systems will impact long-duration space missions by ensuring the health and safety of astronauts. Improved systems for air and water recycling, waste management, and food production will enable missions to operate more sustainably. These advancements reduce the need for resupply from Earth and support the goal of establishing permanent human presence in space.

50. **Question**: What role will international space organizations play in coordinating future exploration efforts?
    Answer: International space organizations will play a crucial role in coordinating future exploration efforts by fostering collaboration, sharing knowledge, and developing standards. Organizations

like the United Nations Office for Outer Space Affairs (UNOOSA) and the International Space Exploration Coordination Group (ISECG) facilitate cooperation among space-faring nations and ensure that exploration activities align with international norms and goals.

51. **Question**: How will future exploration missions address the challenges of space radiation?
Answer: Future exploration missions will address the challenges of space radiation by developing advanced shielding technologies, radiation-hardened electronics, and protective habitats. Research on the ISS and lunar missions will provide data on radiation exposure and inform the design of effective countermeasures. These efforts are critical for ensuring the safety of astronauts on long-duration missions to Mars and beyond.

52. **Question**: What is the significance of commercial partnerships in future space exploration?
Answer: Commercial partnerships are significant in future space exploration as they bring innovation, efficiency, and additional resources to exploration efforts. Companies like SpaceX, Blue Origin, and Boeing are developing technologies and services that support government-led missions. These partnerships reduce costs, increase mission capabilities, and expand access to space.

53. **Question**: How will the experience gained from the ISS influence the design of future space habitats?
Answer: The experience gained from the ISS will influence the design of future space habitats by providing valuable insights into life support, crew dynamics, and habitat operations. Lessons learned from ISS missions will inform the development of sustainable and comfortable living environments for astronauts on the moon, Mars, and beyond. This experience ensures that future habitats are well-equipped to support long-duration missions. (See Figure 5.5.)

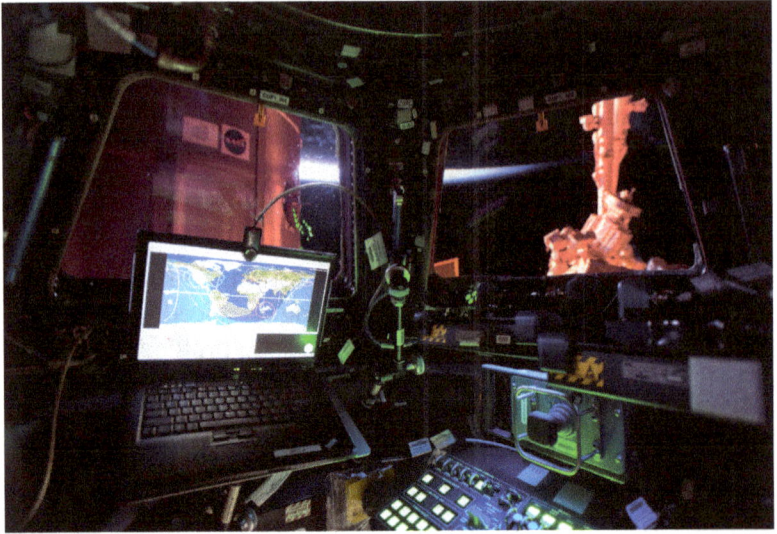

FIGURE 5.5 The experience gained from the ISS will influence the design of future space habitats by providing valuable insights into life support, crew dynamics, and habitat operations. Source: Wikimedia Commons.

54. **Question**: How will international collaboration shape the future of space exploration?

    Answer: International collaboration will shape the future of space exploration by combining the strengths and resources of multiple countries to achieve common goals. Collaborative efforts will enhance scientific discoveries, technological advancements, and the sustainability of exploration missions. The shared vision and cooperative spirit of international partnerships will drive humanity's exploration of the moon, Mars, and beyond.

The International Space Station has played a crucial role in advancing our understanding of space and fostering international cooperation. Through continuous collaboration, space agencies from around the world have contributed to the station's construction, maintenance, and scientific achievements. The ISS has provided a unique microgravity environment for groundbreaking research and technological innovation, significantly impacting fields such as medicine, materials science, and space travel.

As we look toward future exploration efforts on the Moon and Mars, the lessons learned and partnerships formed through the ISS will be crucial in shaping the next era of space exploration. The collaborative spirit and scientific achievements of the ISS provide a strong foundation for humanity's continued journey into the cosmos. The ISS's legacy will be reflected in the success of future missions that build upon its advancements, ensuring that space exploration remains a global endeavor. The continued evolution and success of the ISS underscore the importance of international cooperation in overcoming the challenges of space exploration and achieving our collective goals.

REFERENCES

[Harland05] Harland, David M., *The Story of the International Space Station*, Springer-Praxis, 2005.

[Logsdon96] Logsdon, John M., ed., *Exploring the Unknown: Selected Documents in the History of the U.S. Civil Space Program, Volume II: External Relationships*, NASA SP-4407, 1996.

[Neufeld07] Neufeld, Michael J., *Von Braun: Dreamer of Space, Engineer of War*, Alfred A. Knopf, 2007.

[Newberg19] Newberg, Heidi, and Richard M. Freeland, *The International Space Station: Building for the Future*, Mercury Learning & Information, 2019.

[Seedhouse19] Seedhouse, Erik, *Space Stations: The Art, Science, and Reality of Working in Space*, Springer-Praxis, 2019.

[Smith01] Smith, Marcia S., *The International Space Station: Policy, Plans, and Progress*, Congressional Research Service, Library of Congress, 2001.

[White99] White, Michael, and Brian R. Eddy, *International Space Station: Architecture Beyond Earth*, Smithsonian Institution, 1999.

CHAPTER 6

SPACE FOR ALL: COMMERCIAL VENTURES

INTRODUCTION

The space industry has undergone a significant transformation over the past few decades, driven by the rise of private space companies. These companies have revolutionized space travel and exploration, making it more accessible and affordable. This chapter explores the growth of private space companies, the emergence of space tourism, the potential for asteroid mining, and the broader implications of commercial ventures in space. Major players such as SpaceX, Blue Origin, and Virgin Galactic have made significant contributions, leading to economic and scientific impacts that are reshaping the future of space activities.

As the commercial space sector continues to expand, it promises to democratize access to space, drive technological innovation, and create new economic opportunities. The chapter provides a comprehensive overview of the key developments and trends shaping the future of commercial space activities, emphasizing the collaborative efforts between private companies and government agencies. It also examines the challenges and opportunities in this rapidly evolving industry, including the potential benefits and ethical considerations of space commercialization.

THE RISE OF PRIVATE SPACE COMPANIES

1. **Question**: What factors contributed to the rise of private space companies?

   Answer: Several factors contributed to the rise of private space companies, including advances in technology, increased investment from private sectors, and supportive government policies. The end of the Space Shuttle program in 2011 also opened opportunities for private companies to fill the gap in space transportation. Innovations in reusable rockets, miniaturization of satellite technology, and the vision of entrepreneurs like Elon Musk and Jeff Bezos have propelled the growth of private space ventures.

2. **Question**: How did SpaceX revolutionize the space industry?

   Answer: SpaceX revolutionized the space industry by developing the first commercially-built and operated spacecraft capable of transporting cargo and crew to the International Space Station (ISS). The company pioneered the development of reusable rockets, significantly reducing the cost of launches. Notable milestones include the successful launch and landing of the Falcon 9 rocket, the deployment of the Falcon Heavy, and the Crew Dragon missions. SpaceX's innovations have democratized access to space and inspired a new wave of private space ventures. (See to Figure 6.1.)

FIGURE 6.1 This image features SpaceX's Crew Dragon spacecraft, which is used to trans(t astronauts to and from the International Space Station as part of NASA's Commercial Crew Program. Source: NASA.

3. **Question**: What is the significance of China's commercial space sector growth?
    Answer: China's commercial space sector has seen significant growth with companies like iSpace and OneSpace developing launch vehicles and satellite technologies. This growth is driven by government support and the increasing demand for commercial satellite services.

4. **Question**: What role has Blue Origin played in advancing commercial space travel?
    Answer: Blue Origin, founded by Jeff Bezos, has focused on making space travel more affordable and accessible. The company's New Shepard suborbital rocket has successfully completed multiple flights, carrying both scientific payloads and tourists to the edge of space. Blue Origin is also developing the New Glenn orbital rocket and the Blue Moon lunar lander. The company's emphasis on reusable technology and sustainable space exploration aligns with its long-term vision of enabling millions of people to live and work in space.

5. **Question**: How has Virgin Galactic contributed to the space tourism industry?
    Answer: Virgin Galactic, founded by Richard Branson, has been a pioneer in the space tourism industry. The company's SpaceShipTwo vehicle is designed to carry passengers on suborbital flights, offering a few minutes of weightlessness and spectacular views of Earth. Virgin Galactic has successfully conducted crewed test flights and aims to begin commercial operations soon. The company's efforts have generated significant public interest in space tourism and demonstrated the feasibility of commercial space travel for civilians.

6. **Question**: How has India's Antrix Corporation contributed to commercial space activities?
    Answer: Antrix Corporation, the commercial arm of ISRO, plays a crucial role in promoting India's space capabilities globally. It facilitates satellite launches, transponder leasing, and marketing of space products, boosting India's presence in the global space market.

7. **Question**: What are some other notable private space companies, and what are their contributions?
    Answer: Other notable private space companies include Rocket Lab, Sierra Nevada Corporation, and Relativity Space. Rocket Lab specializes in small satellite launches with its Electron rocket,

providing affordable access to orbit for smaller payloads. Sierra Nevada Corporation is developing the Dream Chaser spaceplane, designed for cargo and crew missions to the ISS. Relativity Space is innovating with 3D-printed rockets, aiming to reduce manufacturing time and costs. These companies contribute to the diversification and growth of the commercial space industry.

8. **Question**: How have government policies supported the growth of private space companies?

   Answer: Government policies have played a crucial role in supporting the growth of private space companies. Initiatives such as NASA's Commercial Orbital Transportation Services (COTS) program and the Commercial Crew Program provided funding and contracts to private companies for developing spacecraft and launch vehicles. Regulatory frameworks, such as the U.S. Commercial Space Launch Competitiveness Act, have encouraged private sector investment and innovation. These policies have facilitated partnerships between public agencies and private companies, driving advancements in space technology.

9. **Question**: What is the significance of reusable rocket technology in commercial space ventures?

   Answer: Reusable rocket technology is significant in commercial space ventures because it drastically reduces the cost of launches by allowing rockets to be flown multiple times. Companies like SpaceX and Blue Origin have demonstrated the feasibility of landing and reusing first-stage boosters, making space missions more economical and sustainable. Reusability also accelerates the frequency of launches, enabling more missions and increasing access to space for a variety of purposes, from scientific research to commercial payloads.

10. **Question**: How has the miniaturization of satellite technology impacted the space industry?

    Answer: The miniaturization of satellite technology has had a profound impact on the space industry by making it more accessible and affordable. Small satellites, or CubeSats, can be developed and launched at a fraction of the cost of traditional satellites. This has opened up space to a wider range of players, including universities, startups, and developing countries. Miniaturized satellites are used for diverse applications such as Earth observation, communications, and scientific research, driving innovation and growth in the space sector.

11. **Question**: What are the achievements of China's private space company iSpace?

   Answer: iSpace became the first private Chinese company to successfully reach orbit with its Hyperbola-1 rocket in 2019. This achievement highlights the potential of China's private space sector to contribute to commercial launch services and satellite deployment.

12. **Question**: What are the economic benefits of commercial space ventures?

   Answer: The economic benefits of commercial space ventures include job creation, technological innovation, and the development of new markets. The growth of private space companies has spurred investment in related industries, such as aerospace manufacturing, software development, and telecommunications. The commercialization of space has also led to the creation of new business opportunities, such as space tourism, satellite services, and asteroid mining, contributing to economic growth and technological advancement.

13. **Question**: How do private space companies contribute to scientific research?

   Answer: Private space companies contribute to scientific research by providing affordable access to space for experiments and observations. Companies like SpaceX and Rocket Lab offer launch services for scientific payloads, enabling researchers to conduct studies in microgravity, observe celestial phenomena, and test new technologies. The involvement of private companies increases the frequency of scientific missions and broadens the scope of research conducted in space.

14. **Question**: How did India's PSLV-C37 mission set a world record?

   Answer: The PSLV-C37 mission, launched by ISRO in 2017, set a world record by deploying 104 satellites in a single launch. This mission demonstrated India's capability in efficiently managing multiple satellite deployments and reinforced its position as a reliable launch service provider.

15. **Question**: What challenges do private space companies face in their operations?

   Answer: Private space companies face several challenges, including high development and operational costs, regulatory hurdles, and technical risks. The cost of developing and launching spacecraft remains significant, requiring substantial investment and careful financial planning. Regulatory requirements, such as licensing and safety standards, can be complex and time-consuming. Technical challenges, such as ensuring the reliability and safety of reusable rockets, also pose risks to the success of missions.

16. **Question**: How has the partnership between NASA and private companies evolved?

    Answer: The partnership between NASA and private companies has evolved from traditional contracting to more collaborative and innovative approaches. NASA's Commercial Crew and Cargo Programs have fostered public–private partnerships, where private companies develop and operate spacecraft with funding and oversight from NASA. This approach has led to successful missions, such as SpaceX's Dragon cargo and crew flights to the ISS. The collaboration has expanded to include lunar exploration and Mars missions, leveraging the strengths of both NASA and private industry. (See Figure 6.2.)

FIGURE 6.2 The partnership between NASA and private companies has evolved from traditional contracting to more collaborative and innovative approaches. NASA's Commercial Crew and Cargo Programs have fostered public–private partnerships, where private companies develop and operate spacecraft with funding and oversight from NASA. Source: Wikimedia Commons.

17. **Question**: What is the role of venture capital in the growth of private space companies?

    Answer: Venture capital plays a crucial role in the growth of private space companies by providing the necessary funding to develop new technologies and scale operations. Investment from venture capital firms enables startups to innovate, conduct research and development, and bring their products to market. High-profile investments in companies like SpaceX, Rocket Lab, and Planet Labs have demonstrated the potential for significant returns, attracting more venture capital to the space sector.

18. **Question**: How do private space companies contribute to global connectivity?

    Answer: Private space companies contribute to global connectivity by developing and deploying satellite constellations for communication and Internet services. Companies like SpaceX (with Starlink), OneWeb, and Amazon (with Project Kuiper) are working to provide high-speed Internet access to underserved and remote areas worldwide. These initiatives aim to bridge the digital divide, enhance global communication, and support economic development through improved connectivity.

19. **Question**: What are the environmental considerations for commercial space activities?

    Answer: Environmental considerations for commercial space activities include the impact of rocket launches on the atmosphere, space debris, and sustainable use of space resources. Rocket launches produce emissions that can affect the upper atmosphere, and the increasing number of satellites raises concerns about space debris. Companies and regulatory bodies are working on solutions to mitigate these impacts, such as developing cleaner propulsion technologies, designing satellites for safe deorbiting, and implementing guidelines for responsible space activities.

TOURISM, ASTEROID MINING, AND BEYOND

20. **Question**: What is space tourism, and how has it developed?

    Answer: Space tourism refers to commercial activities that allow private individuals to travel to space for leisure or adventure. It has developed through the efforts of companies like Virgin Galactic, Blue Origin, and SpaceX. Virgin Galactic's SpaceShipTwo and Blue Origin's New Shepard offer suborbital flights that provide a few minutes of weightlessness and stunning views of Earth. SpaceX plans to offer orbital tourism with its Crew Dragon spacecraft. Space tourism is still in its early stages, but it has generated significant interest and investment.

21. **Question**: What are the potential benefits of space tourism?
    Answer: The potential benefits of space tourism include increased public interest in space exploration, economic growth, and technological innovation. Space tourism can inspire a new generation of space enthusiasts and support educational initiatives. The industry also creates jobs and stimulates economic activity in related sectors. Technological advancements developed for space tourism, such as reusable rockets and life support systems, can have broader applications in other space missions and industries.

22. **Question**: What challenges must be addressed for space tourism to become mainstream?
    Answer: Challenges for mainstream space tourism include ensuring the safety and reliability of spaceflights, reducing costs, and addressing regulatory and environmental concerns. Safety is paramount, and companies must rigorously test their vehicles and systems to protect passengers. Reducing costs is essential to make space tourism accessible to a broader audience. Regulatory frameworks must evolve to accommodate commercial space tourism, and environmental impacts must be managed responsibly.

23. **Question**: How does asteroid mining work, and what are its potential benefits?
    Answer: Asteroid mining involves extracting valuable materials from asteroids, such as metals, water, and other resources. Companies like Planetary Resources and Deep Space Industries have explored the potential of mining asteroids for resources that could support space missions or be brought back to Earth. The potential benefits of asteroid mining include access to abundant raw materials, reducing the cost of space missions by providing in-situ resources, and supporting the development of space infrastructure.

24. **Question**: What technologies are needed for successful asteroid mining?
    Answer: Successful asteroid mining requires several advanced technologies, including robotic spacecraft for prospecting and mining, advanced sensors for resource detection, and processing equipment to extract and refine materials. Autonomous systems and artificial intelligence are crucial for operating in the harsh and remote environment of space. Efficient propulsion systems are needed for transportation to and from asteroids. Developing these technologies involves significant research and investment.

25. **Question**: What are the economic implications of asteroid mining?
Answer: The economic implications of asteroid mining are substantial, with the potential to create new industries and markets. Access to rare and valuable materials could drive technological advancements and economic growth. In-space resource utilization can reduce the cost of space missions and support the development of space habitats and infrastructure. However, significant challenges must be overcome, including high initial investment costs and regulatory hurdles.

26. **Question**: How does international law address the exploitation of space resources?
Answer: International law, through treaties like the Outer Space Treaty of 1967 and the Moon Agreement, provides a framework for the exploration and use of space resources. These treaties emphasize that space is the province of all humankind and that activities should benefit all countries. However, the legal framework for commercial exploitation of space resources is still evolving. Countries like the United States have enacted national laws to support commercial space activities, but international consensus is needed to ensure fair and sustainable practices.

27. **Question**: What role do public–private partnerships play in advancing space commercialization?
Answer: Public–private partnerships play a crucial role in advancing space commercialization by combining the resources, expertise, and innovation of both sectors. These partnerships enable the development of new technologies, share risks and costs, and accelerate the implementation of space missions. Examples include NASA's partnerships with SpaceX and Boeing for crewed missions and collaborations on lunar exploration initiatives. Such partnerships enhance the capabilities and sustainability of space exploration.

28. **Question**: How have advances in materials science impacted commercial space ventures?
Answer: Advances in materials science have significantly impacted commercial space ventures by improving the performance, durability, and cost-effectiveness of spacecraft and equipment. New materials, such as advanced composites and lightweight alloys, enhance the structural integrity of rockets and satellites. Innovations in thermal protection materials improve the safety of reentry vehicles. These advancements enable more ambitious missions and reduce the overall costs of space operations.

29. **Question**: What is the significance of satellite mega-constellations in the commercial space sector?

    Answer: Satellite mega-constellations, consisting of hundreds or thousands of small satellites in low Earth orbit, are significant for providing global Internet coverage and enhancing communication networks. Companies like SpaceX (Starlink) and OneWeb are developing these constellations to offer high-speed Internet access to remote and underserved regions. Mega-constellations can also support applications in Earth observation, navigation, and disaster response. Their deployment represents a major shift in the commercial space sector, increasing connectivity and creating new business opportunities. (See Figure 6.3.)

FIGURE 6.3 The Starlink satellite constellation, developed by SpaceX, is an example of a satellite mega-constellation designed to provide global internet coverage. Source: Wikimedia Commons.

30. **Question**: What are the environmental concerns associated with satellite mega-constellations?

    Answer: Environmental concerns associated with satellite mega-constellations include the potential for increased space debris, light pollution, and the impact on astronomical observations. The large number of satellites increases the risk of collisions and contributes to the growing problem of space debris. Light pollution from satellites can interfere with astronomical observations and research. Companies and regulatory bodies are working on measures to mitigate these impacts, such as improving satellite design for deorbiting and implementing guidelines for responsible satellite operations.

31. Question: How has the concept of space habitats evolved in commercial ventures?
Answer: The concept of space habitats has evolved from theoretical designs to practical development efforts by companies like Bigelow Aerospace and Orbital Assembly Corporation. These companies are working on expandable habitats and rotating space stations that provide artificial gravity. Space habitats aim to support long-term human presence in space for purposes such as research, tourism, and industrial activities. The development of space habitats represents a significant step toward the commercialization of space and the establishment of permanent human settlements beyond Earth.

32. Question: What are the potential benefits of in-space manufacturing?
Answer: In-space manufacturing offers potential benefits such as reducing the cost and complexity of space missions, enabling the production of materials and equipment on demand, and supporting long-duration missions. Technologies like 3D printing can produce spare parts, tools, and even habitat components in space, reducing the need for resupply from Earth. In-space manufacturing can also create new materials with unique properties that are difficult or impossible to produce in Earth's gravity.

33. Question: How is the concept of space mining different from traditional mining on Earth?
Answer: Space mining differs from traditional mining on Earth in several key aspects, including the environment, technology, and logistics. Space mining operations must contend with microgravity, extreme temperatures, and radiation. Robotic systems and autonomous technologies are essential for extracting resources in these harsh conditions. The logistics of transporting mined materials to space habitats or back to Earth also pose significant challenges. Space mining requires innovative approaches to resource extraction and utilization.

34. Question: What are the ethical considerations of commercial space ventures?
Answer: Ethical considerations of commercial space ventures include ensuring the responsible use of space resources, preventing the militarization of space, and promoting equitable access to space benefits. Companies must consider the long-term sustainability of their activities and avoid contributing to space debris.

There are also concerns about the potential for space exploration to exacerbate inequalities between countries. Ethical frameworks and international cooperation are essential to address these issues and ensure that space commercialization benefits all of humanity.

35. **Question**: How do space startups contribute to innovation in the space industry?
    Answer: Space startups contribute to innovation in the space industry by introducing new technologies, business models, and approaches to space exploration. These companies often focus on niche markets or develop disruptive technologies that challenge traditional aerospace practices. Examples include CubeSat developers, propulsion system innovators, and companies working on space debris removal. The agility and creativity of startups drive technological advancements and expand the scope of commercial space activities.

36. **Question**: What is the role of government agencies in regulating commercial space activities?
    Answer: Government agencies play a crucial role in regulating commercial space activities to ensure safety, security, and compliance with international agreements. Agencies like the Federal Aviation Administration (FAA) in the United States oversee the licensing of space launches and reentries, while the Federal Communications Commission (FCC) regulates satellite communications. These agencies set standards for environmental protection, space traffic management, and the prevention of space debris. Effective regulation supports the sustainable growth of the commercial space sector.

37. **Question**: How has the market for satellite data services evolved?
    Answer: The market for satellite data services has evolved rapidly, driven by advances in satellite technology, increased demand for real-time data, and the growth of applications in various industries. Satellite data services provide valuable information for agriculture, disaster response, environmental monitoring, and urban planning. Companies like Planet Labs and Maxar Technologies offer high-resolution Earth observation data, while others focus on weather forecasting, maritime tracking, and remote sensing. The increasing availability of satellite data supports decision-making and drives economic growth.

38. **Question**: What are the challenges of ensuring the long-term sustainability of commercial space activities?

    Answer: Ensuring the long-term sustainability of commercial space activities involves addressing challenges such as space debris management, environmental impacts, and resource utilization. Effective space traffic management is needed to prevent collisions and reduce the risk of debris generation. Developing environmentally friendly propulsion systems and practices can minimize the impact of launches on the atmosphere. Sustainable resource utilization strategies, such as in-situ resource utilization (ISRU), are essential for long-duration missions and the establishment of space infrastructure.

39. **Question**: How does international collaboration enhance commercial space ventures?

    Answer: International collaboration enhances commercial space ventures by pooling resources, sharing expertise, and fostering innovation. Collaborative efforts can reduce costs, spread risks, and accelerate the development of new technologies. International partnerships also facilitate access to global markets and regulatory frameworks, enabling companies to operate more effectively. Examples include joint satellite missions, cross-border investments, and collaborative research initiatives. Such collaborations support the growth and sustainability of the commercial space sector.

40. **Question**: What is the potential impact of space-based solar power on the commercial space industry?

    Answer: Space-based solar power has the potential to revolutionize the commercial space industry by providing a reliable and sustainable energy source. Solar power satellites can collect solar energy in space and transmit it to Earth using microwave or laser technology. This approach offers continuous energy production, unaffected by weather or time of day. The development of space-based solar power systems could drive advancements in space infrastructure, reduce dependence on fossil fuels, and support global energy needs. However, significant technical and economic challenges must be overcome to realize this potential. (See Figure 6.4.)

FIGURE 6.4 The concept of lunar solar power involves using solar cells and microwave transmitters to collect and transmit solar energy from the Moon to Earth. Source: Wikimedia Commons.

41. **Question**: How have advancements in propulsion technology influenced commercial space missions?

   Answer: Advancements in propulsion technology have significantly influenced commercial space missions by improving efficiency, reducing costs, and enabling new mission profiles. Innovations such as electric propulsion, reusable rocket engines, and advanced chemical propulsion systems have enhanced the capabilities of launch vehicles and spacecraft. These advancements support a wide range of commercial activities, from satellite deployment to deep space exploration. Improved propulsion technology enables more frequent and diverse missions, driving growth in the commercial space sector.

42. **Question**: What is the role of spaceports in supporting commercial space activities?

    Answer: Spaceports play a vital role in supporting commercial space activities by providing infrastructure for rocket launches, vehicle recovery, and payload integration. They serve as hubs for space transportation, facilitating access to space for a variety of missions. Spaceports like Kennedy Space Center, Vandenberg Space Force Base, and commercial facilities like Spaceport America offer launch services for government, commercial, and scientific missions. The development of new spaceports around the world enhances global access to space and supports the growth of the commercial space industry.

43. **Question**: How do space insurance and risk management support commercial space ventures?

    Answer: Space insurance and risk management are essential for supporting commercial space ventures by mitigating financial risks associated with space missions. Insurance policies cover various aspects of space activities, including launch, in-orbit operations, and satellite deployment. Effective risk management practices help companies identify and address potential hazards, ensuring the safety and success of their missions. The availability of space insurance provides financial security and encourages investment in commercial space ventures.

44. **Question**: What are the potential applications of artificial intelligence (AI) in commercial space activities?

    Answer: Artificial intelligence (AI) has numerous potential applications in commercial space activities, including autonomous spacecraft navigation, data analysis, and mission planning. AI algorithms can optimize flight trajectories, monitor spacecraft health, and process large volumes of scientific data. AI-driven systems enhance the efficiency and reliability of space missions, enabling more complex and ambitious projects. The integration of AI into space operations supports innovation and growth in the commercial space sector.

45. **Question**: How has the development of small launch vehicles impacted the commercial space industry?

    Answer: The development of small launch vehicles has impacted the commercial space industry by providing affordable and flexible access to space for small satellites. Companies like Rocket Lab, Virgin Orbit, and Astra have developed dedicated small launch vehicles that cater to the growing demand for small satellite deployment. These vehicles offer more frequent and tailored launch opportunities, reducing reliance on rideshare missions with larger rockets. The availability of small launch vehicles supports the expansion of the small satellite market and drives innovation in space applications.

46. **Question**: What are the benefits of space-based manufacturing for commercial ventures?

    Answer: Space-based manufacturing offers several benefits for commercial ventures, including the ability to produce high-quality materials and components in microgravity. Manufacturing processes in space can create products with unique properties, such as stronger materials, more precise structures, and improved performance. Space-based manufacturing can also reduce the need for resupply from Earth, supporting long-duration missions and the development of space habitats. This capability enhances the sustainability and economic viability of space exploration. (See Figure 6.5.)

FIGURE 6.5 The OneWeb satellite manufacturing facility, located in Merritt Island, Florida, demonstrates advancements in space-based manufacturing.
Source: Wikimedia Commons.

47. **Question**: How do public outreach and education initiatives support commercial space ventures?

    Answer: Public outreach and education initiatives support commercial space ventures by raising awareness, generating interest, and inspiring the next generation of space professionals. Companies like SpaceX, Blue Origin, and Virgin Galactic engage with the public through media coverage, educational programs, and outreach events. These initiatives highlight the achievements and potential of commercial space activities, fostering a positive perception and encouraging investment. Education programs also help develop the skilled workforce needed for the future growth of the space industry.

48. **Question**: What are the potential impacts of space commercialization on global economies?

    Answer: The potential impacts of space commercialization on global economies include the creation of new industries, job opportunities, and technological advancements. The commercialization of space drives innovation in various sectors, such as aerospace, telecommunications, and materials science. It also stimulates economic growth by attracting investment and fostering international trade. The development of space resources and infrastructure can support sustainable economic development and improve the quality of life on Earth.

49. **Question**: How do regulatory frameworks support the responsible development of commercial space activities?

    Answer: Regulatory frameworks support the responsible development of commercial space activities by establishing standards for safety, environmental protection, and resource management. Agencies like the FAA, FCC, and international bodies set guidelines for licensing, space traffic management, and debris mitigation. These regulations ensure that commercial space ventures operate safely and sustainably, protecting both space assets and the broader environment. Effective regulatory frameworks promote innovation while safeguarding public and environmental interests.

50. **Question**: What role do international agreements play in the commercialization of space?

    Answer: International agreements play a crucial role in the commercialization of space by providing a framework for cooperation, resource sharing, and conflict resolution. Treaties like the Outer Space Treaty and the Moon Agreement set principles for the peaceful use of space and the equitable distribution of benefits. These agreements encourage collaboration among countries and private entities, fostering a stable and predictable environment for commercial space activities. International cooperation ensures that space commercialization advances shared goals and benefits all of humanity.

51. **Question**: How has the concept of space mining evolved over time?

    Answer: The concept of space mining has evolved from theoretical proposals to practical development efforts, driven by advances in technology and increasing interest from private companies. Early concepts focused on the potential for extracting valuable resources from the moon and asteroids. Recent developments include the identification of target asteroids, the design of mining spacecraft, and the establishment of legal frameworks for resource ownership. The evolution of space mining reflects growing recognition of its potential to support space exploration and economic development.

52. **Question**: What are the key challenges in developing sustainable commercial space activities?
    Answer: Key challenges in developing sustainable commercial space activities include managing space debris, ensuring equitable access to space resources, and minimizing environmental impacts. Effective space traffic management and debris mitigation strategies are essential to prevent collisions and maintain a safe space environment. Sustainable practices for resource utilization and waste management are critical for long-term exploration. Addressing these challenges requires collaboration among governments, industry, and international organizations to establish standards and best practices.

53. **Question**: How do commercial space ventures contribute to the advancement of space science?
    Answer: Commercial space ventures contribute to the advancement of space science by providing new opportunities for research and exploration. Companies like SpaceX, Rocket Lab, and Planet Labs offer launch services and satellite platforms for scientific missions. These ventures enable more frequent and diverse experiments, enhancing our understanding of space phenomena. Commercial innovations in spacecraft design, instrumentation, and data processing support the scientific community and drive discoveries in astronomy, planetary science, and other fields.

54. **Question**: What is the future outlook for commercial space ventures?
    Answer: The future outlook for commercial space ventures is promising, with continued growth and innovation expected in various sectors. Advances in technology, increasing investment, and supportive policies will drive the expansion of space tourism, satellite services, and resource utilization. Emerging markets, such as space habitats and in-space manufacturing, will create new opportunities and challenges. The collaboration between private companies and government agencies will enhance the capabilities and sustainability of space exploration. As the commercial space industry evolves, it will play a crucial role in shaping humanity's future in space.

The rise of private space companies has fundamentally transformed the landscape of space exploration and commercialization. From pioneering reusable rockets to enabling space tourism and asteroid mining, these ventures are opening new frontiers and creating unprecedented opportunities. Advancements made by companies like SpaceX, Blue Origin, and Virgin Galactic have not only reduced the cost of space access but also inspired a new generation of entrepreneurs and innovators.

Looking to the future, the continued growth and innovation in commercial space activities promise to make space more accessible and beneficial for all of humanity. The collaboration between private companies and government agencies will be crucial in addressing challenges such as space debris, regulatory frameworks, and sustainable practices. The economic benefits of commercial space ventures, including job creation, technological innovation, and new markets, will drive further investment and development in the industry. Lessons learned and technologies developed through these commercial ventures will also support future scientific missions and space exploration efforts, ensuring that the benefits of space activities are shared globally.

REFERENCES

[AndersonRide21] Anderson, Chris, and Sally Ride, *Space Entrepreneurs: How Elon Musk, Jeff Bezos, and Richard Branson Are Changing the World*, Harper Business, 2021.

[Cadbury06] Cadbury, Deborah, *Space Race: The Epic Battle Between America and the Soviet Union for Dominion of Space*, HarperCollins, 2006.

[Chaikin94] Chaikin, Andrew, *A Man on the Moon: The Voyages of the Apollo Astronauts*, Penguin Books, 1994.

[Neufeld07] Neufeld, Michael J., *Von Braun: Dreamer of Space, Engineer of War*, Alfred A. Knopf, 2007.

[Oberg99] Oberg, James E., *Space Power Theory*, Government Printing Office, 1999.

[Seedhouse19] Seedhouse, Erik, *SpaceX: Making Commercial Spaceflight a Reality*, Springer-Praxis, 2019.

[Winter05] Winter, Frank H., and William P. Barry, *Rockets and People: Volume I*, NASA History Series SP-2005-4110, 2005.

CHAPTER 7

CHALLENGES AND FUTURE PROSPECTS

INTRODUCTION

Space exploration faces numerous challenges and presents significant opportunities for future advancements. This chapter delves into the critical hurdles that must be overcome to ensure sustainable and ethical space exploration, focusing on space debris, radiation, and ethical considerations. It examines the principles of sustainable exploration and habitation, highlighting the importance of minimizing environmental impact, ensuring long-term resource availability, and fostering international cooperation. Additionally, the chapter explores the quest for interstellar travel, discussing the technological advancements and theoretical concepts that may one day make journeys to other star systems possible. By addressing these challenges and prospects, the chapter provides a comprehensive overview of the factors shaping the future of space exploration.

SPACE DEBRIS, RADIATION, AND ETHICAL CONSIDERATIONS

1. **Question**: What is space debris, and why is it a concern?
   Answer: Space debris, also known as space junk, consists of defunct satellites, spent rocket stages, and fragments from collisions and disintegrations. It poses a significant threat to operational satellites, spacecraft, and the International Space Station (ISS). Collisions with space debris can cause damage, create more

debris, and potentially endanger human lives. The growing amount of debris increases the likelihood of collisions, which can lead to a cascade effect known as the Kessler Syndrome, rendering certain orbits unusable. (See Figure 7.1.)

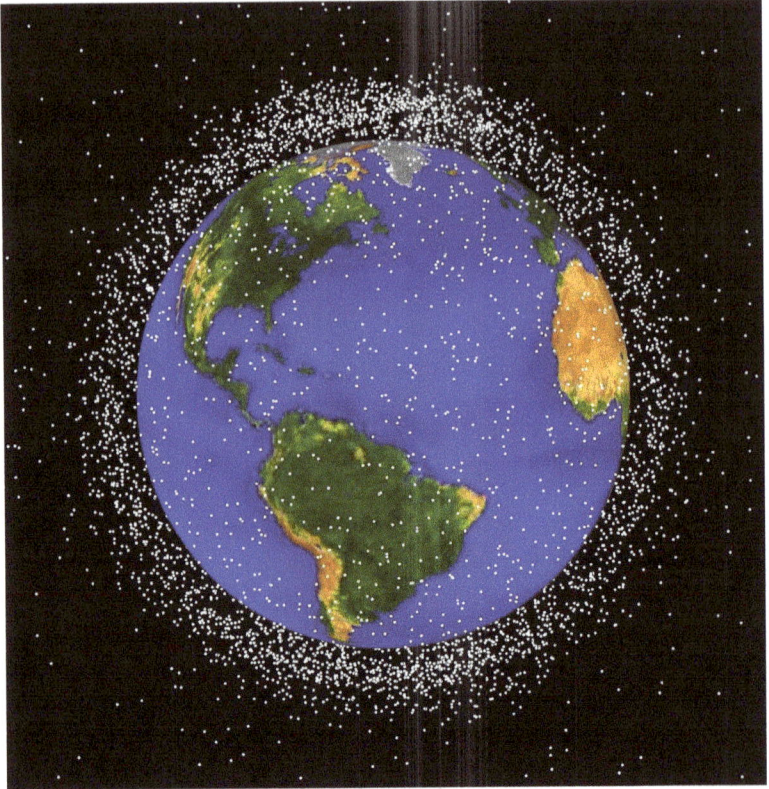

FIGURE 7.1 An artist's rendition of Earth surrounded by a field of space debris, highlighting the growing issue of orbital debris and its potential impact on space missions. Source: Wikimedia Commons.

2. **Question**: How is space debris tracked and managed?
   Answer: Space debris is tracked and managed using ground-based radar and optical telescopes, as well as space-based sensors. Organizations like the United States Space Surveillance Network (SSN) monitor and catalog objects in orbit. Collision avoidance maneuvers are conducted to protect spacecraft from potential impacts. Efforts to mitigate space debris include designing satellites to deorbit at the end of their operational life, using technologies to capture and remove debris, and adhering to international guidelines for space operations.

3. **Question**: What technological solutions are being developed to address space debris?

   Answer: Technological solutions to address space debris include active debris removal (ADR) systems, such as robotic arms, nets, harpoons, and drag sails. ADR missions aim to capture and deorbit large debris objects. Other approaches include using ground-based lasers to nudge debris into lower orbits where it will burn up in the atmosphere. Developing satellite servicing and refueling capabilities can also extend the operational life of satellites, reducing the creation of new debris.

4. **Question**: What is the impact of radiation on space missions?

   Answer: Radiation poses a significant challenge for space missions, affecting both human health and electronic systems. Cosmic rays and solar particle events (SPEs) can damage DNA, increase cancer risks, and cause acute radiation sickness in astronauts. Radiation can also degrade spacecraft electronics, leading to malfunctions and mission failures. Protecting against radiation is crucial for long-duration missions beyond low Earth orbit, such as those to Mars and deep space.

5. **Question**: What measures is China taking to address space debris?

   Answer: China is actively developing technologies for space debris mitigation, including plans for robotic cleanup missions and the deployment of satellites equipped with nets or harpoons to capture debris. These efforts aim to reduce the risk of collisions and maintain sustainable space operations.

6. **Question**: How do spacecraft protect against radiation?

   Answer: Spacecraft protect against radiation using shielding materials, such as polyethylene and water, which can absorb and block harmful particles. Advanced materials, like hydrogen-rich composites, offer improved protection. Spacecraft design also includes radiation-hardened electronics and the use of redundant systems to ensure reliability. For human missions, habitat designs incorporate storm shelters where astronauts can take refuge during solar particle events.

7. **Question**: What ethical considerations are involved in space exploration?

   Answer: Ethical considerations in space exploration include the responsible use of space resources, the preservation of space environments, and the equitable distribution of benefits. Issues such as space debris management, planetary protection, and the potential impact on extraterrestrial life must be addressed. Ethical frameworks should guide decision-making to ensure that space activities do not harm the environment or future generations and that the benefits of space exploration are shared globally.

8. **Question**: What is planetary protection, and why is it important?
    Answer: Planetary protection involves preventing biological contamination of both Earth and other celestial bodies. It is important to avoid introducing Earth microbes to other planets, which could interfere with the search for extraterrestrial life and disrupt pristine environments. Conversely, it also aims to prevent the return of potentially harmful extraterrestrial organisms to Earth. Adhering to planetary protection protocols ensures the integrity of scientific research and the safety of Earth's biosphere.

9. **Question**: How is India addressing the issue of space debris?
    Answer: India is implementing measures to minimize space debris through stringent guidelines for satellite launches and end-of-life disposal. ISRO is also exploring technologies for active debris removal and developing strategies for tracking and managing debris in orbit.

10. **Question**: How is India addressing the issue of radiation protection for its space missions?
    Answer: ISRO is focusing on developing better shielding technologies and radiation-hardened electronics to protect its spacecraft and future astronauts. Initiatives include studying the effects of space radiation and developing materials that can provide effective protection against cosmic rays and solar particles.

11. **Question**: How do international agreements address ethical considerations in space exploration?
    Answer: International agreements, such as the Outer Space Treaty of 1967 and the Moon Agreement of 1984, provide a legal framework for space exploration. These treaties emphasize that space activities should benefit all humankind, prohibit the appropriation of celestial bodies, and promote the peaceful use of space. They also address issues like noncontamination, resource utilization, and the avoidance of harmful interference. These agreements guide ethical conduct and international cooperation in space.

12. **Question**: What role does public opinion play in shaping space exploration policies?

    Answer: Public opinion plays a significant role in shaping space exploration policies by influencing government priorities and funding decisions. Public support for space missions can drive investments in space programs and inspire political leaders to pursue ambitious exploration goals. Engaging the public through education, outreach, and transparent communication helps build consensus on the ethical and sustainable conduct of space activities.

13. **Question**: What advancements has China made in radiation protection for space missions?

    Answer: China is developing advanced radiation shielding materials and technologies to protect astronauts on long-duration missions. The Tiangong space station includes enhanced radiation protection features to ensure the safety of its crew. (See Figure 7.2.)

FIGURE 7.2 China is developing advanced radiation shielding materials and technologies to protect astronauts on long-duration missions. The Tiangong space station includes enhanced radiation protection features to ensure the safety of its crew. Source: Wikimedia Commons.

14. **Question**: How can space exploration be made more inclusive and equitable?

    Answer: Space exploration can be made more inclusive and equitable by ensuring diverse representation in space agencies, fostering international collaboration, and making space benefits accessible to all countries. Encouraging participation from under-represented groups in STEM fields and providing opportunities for global collaboration on space projects can help achieve this goal. Policies and programs that support inclusive education, outreach, and capacity-building in space sciences are essential.

SUSTAINABLE EXPLORATION AND HABITATION

15. **Question**: What principles guide sustainable space exploration?

    Answer: Principles guiding sustainable space exploration include minimizing environmental impact, ensuring long-term resource availability, and fostering international cooperation. Sustainable exploration practices aim to reduce space debris, protect planetary environments, and use in-situ resources efficiently. Collaboration among spacefaring nations and adherence to international guidelines help ensure that exploration activities are conducted responsibly and benefit all humankind.

16. **Question**: How does in-situ resource utilization (ISRU) support sustainable space exploration?

    Answer: In-situ resource utilization (ISRU) supports sustainable space exploration by enabling the use of local resources, such as water, minerals, and regolith, to support human missions. ISRU technologies can extract and process these resources to produce essentials like water, oxygen, fuel, and building materials. This approach reduces the need for resupply from Earth, lowers mission costs, and supports long-duration exploration and habitation.

17. **Question**: What technologies are being developed for sustainable lunar exploration?

    Answer: Technologies for sustainable lunar exploration include lunar rovers for resource prospecting, ISRU systems for extracting water and oxygen from lunar regolith, and 3D printing for building habitats. Advanced life support systems, radiation shielding, and energy storage solutions are also being developed. These technologies aim to establish a permanent human presence on the moon, supporting scientific research and preparing for future missions to Mars.

18. **Question**: How can habitats be designed for sustainable long-term habitation on the moon and Mars?

    Answer: Habitats for sustainable long-term habitation on the moon and Mars must incorporate advanced life support systems, radiation protection, and efficient resource utilization. Modular and expandable designs, using local materials and 3D printing, can create robust and adaptable living spaces. Habitats should provide reliable air, water, and food supplies, manage waste effectively, and ensure crew safety and well-being. Incorporating renewable energy sources, such as solar power, is also essential. (See Figure 7.3.)

FIGURE 7.3 Lunar surface habitats designed for sustainable long-term habitation must incorporate advanced life support systems, radiation protection, and efficient resource utilization. Source: Wikimedia Commons.

19. **Question**: What are the challenges of establishing a sustainable human presence on Mars?
Answer: Challenges of establishing a sustainable human presence on Mars include ensuring reliable life support, managing radiation exposure, and dealing with the planet's harsh environment. Mars' thin atmosphere, extreme temperatures, and dust storms pose significant risks. Developing technologies for ISRU, habitat construction, and energy production is crucial. Additionally, psychological and social factors must be addressed to support the mental health and well-being of crews on long-duration missions.

20. **Question**: How does the concept of a circular economy apply to space exploration?
Answer: The concept of a circular economy applies to space exploration by emphasizing the reuse, recycling, and efficient use of resources. In space habitats, closed-loop life support systems can recycle air, water, and waste. Using modular and upgradable designs for spacecraft and habitats reduces material waste. ISRU technologies further support a circular economy by utilizing local resources. This approach enhances sustainability and reduces the need for resupply from Earth.

21. **Question**: What role do renewable energy sources play in sustainable space exploration?
Answer: Renewable energy sources, such as solar and nuclear power, play a critical role in sustainable space exploration by providing reliable and long-term energy for habitats, rovers, and scientific instruments. Solar power is particularly valuable on the moon and Mars, where sunlight is abundant. Nuclear power offers a consistent energy supply, especially in environments with limited sunlight. Using renewable energy reduces reliance on Earth-based resources and supports continuous operations.

22. **Question**: How can waste management be improved in space habitats?
Answer: Waste management in space habitats can be improved through the development of closed-loop systems that recycle waste into useful products. Technologies like bioreactors and composting units can convert organic waste into fertilizers for growing food. Recycling systems can process materials like plastics and metals into new components using 3D printing. Efficient waste management reduces the need for resupply, supports sustainability, and enhances the habitability of space environments.

23. **Question**: What is the significance of international cooperation in sustainable space exploration?
    Answer: International cooperation is significant in sustainable space exploration as it allows for the sharing of resources, expertise, and technologies. Collaborative efforts, such as the International Space Station (ISS) and the Artemis program, demonstrate the benefits of working together to achieve common goals. Cooperation helps distribute costs, reduces redundancy, and enhances the capabilities of space missions. It also promotes peaceful use of space and ensures that exploration benefits all humankind.

24. **Question**: How does sustainable space exploration contribute to Earth-based sustainability efforts?
    Answer: Sustainable space exploration contributes to Earth-based sustainability efforts by driving technological innovations and providing valuable data for environmental monitoring. Technologies developed for closed-loop life support, renewable energy, and efficient resource utilization have applications on Earth. Satellite observations from space support climate science, natural resource management, and disaster response. The principles and practices of sustainable space exploration can inform and inspire sustainability initiatives on Earth.

THE QUEST FOR INTERSTELLAR TRAVEL

25. **Question**: What are the main challenges of interstellar travel?
    Answer: The main challenges of interstellar travel include the vast distances, high speeds required, and the need for sustainable life support systems. Traveling to even the nearest star, Proxima Centauri, would take thousands of years with current propulsion technologies. Achieving sufficient speed while ensuring the safety and sustainability of the spacecraft and crew presents significant technical and engineering challenges. Additionally, the mission duration and isolation pose psychological and social challenges for the crew.

26. **Question**: What propulsion technologies are being considered for interstellar travel?
    Answer: Propulsion technologies being considered for interstellar travel include nuclear propulsion (thermal and electric), ion propulsion, solar sails, and antimatter propulsion. Nuclear propulsion offers high efficiency and thrust, while ion propulsion

provides continuous low-thrust acceleration. Solar sails harness the pressure of sunlight for propulsion, and antimatter propulsion promises extremely high energy density. These technologies aim to achieve the high speeds necessary for interstellar travel.

27. **Question**: How could nuclear propulsion benefit interstellar missions?

    Answer: Nuclear propulsion could benefit interstellar missions by providing a powerful and efficient means of propulsion. Nuclear thermal rockets use nuclear reactions to heat propellant, generating high thrust. Nuclear electric propulsion uses nuclear reactors to generate electricity for ion thrusters, offering sustained acceleration over long durations. These technologies can significantly reduce travel time to distant stars, making interstellar missions more feasible.

28. **Question**: What are solar sails, and how do they work?

    Answer: Solar sails are a form of spacecraft propulsion that uses large, reflective sails to capture the momentum of photons from sunlight. As photons strike the sail, they transfer momentum, generating thrust. Solar sails do not require propellant, making them ideal for long-duration missions. They can achieve continuous acceleration, gradually increasing their speed over time. Solar sails are being developed for missions to explore the outer solar system and beyond.

29. **Question**: What is antimatter propulsion, and what are its potential advantages?

    Answer: Antimatter propulsion is a theoretical propulsion method that uses the annihilation of matter and antimatter to produce energy. When antimatter particles collide with matter, they annihilate each other, releasing vast amounts of energy. This energy can be used to generate thrust. Antimatter propulsion offers extremely high energy density, potentially enabling spacecraft to achieve near-light speeds. However, producing and storing antimatter in sufficient quantities remains a significant challenge.

30. **Question**: How does the Breakthrough Starshot initiative aim to achieve interstellar travel?

    Answer: The Breakthrough Starshot initiative aims to achieve interstellar travel by developing small, lightweight spacecraft (called StarChips) propelled by powerful ground-based lasers. The StarChips would be equipped with solar sails and accelerated to a significant fraction of the speed of light. The goal is to reach the Alpha Centauri system within twenty years of launch. This innovative approach leverages advances in miniaturization, photonics, and materials science to overcome the challenges of interstellar travel.

31. **Question**: What role does artificial intelligence (AI) play in interstellar missions?

    Answer: AI plays a crucial role in interstellar missions by enabling autonomous operations, data analysis, and decision-making. AI systems can manage spacecraft functions, perform scientific research, and adapt to unforeseen circumstances without real-time human intervention. The vast distances involved in interstellar travel necessitate AI-driven systems to ensure mission success and handle the complexities of deep space exploration.

32. **Question**: How can sustainable life support systems be developed for interstellar travel?

    Answer: Sustainable life support systems for interstellar travel must be highly efficient and capable of recycling air, water, and waste. Closed-loop systems that utilize bioregenerative processes, such as growing plants for food and oxygen, are essential. Advanced recycling technologies can convert waste into usable resources. These systems must be robust and reliable, capable of supporting crews for extended durations without resupply from Earth.

33. **Question**: What are the psychological and social challenges of long-duration interstellar missions?

    Answer: The psychological and social challenges of long-duration interstellar missions include isolation, confinement, and the stress of prolonged space travel. Crews must cope with being away from Earth for extended periods, possibly spanning generations. Mental health support, effective communication systems, and social activities are essential to maintain crew morale and cohesion. Psychological resilience and team dynamics play a critical role in the success of such missions.

34. **Question**: What ethical considerations must be addressed for interstellar travel?

    Answer: Ethical considerations for interstellar travel include the potential impact on extraterrestrial environments, the rights and welfare of crew members, and the long-term implications for humanity. Ensuring that missions do not harm alien ecosystems or violate planetary protection principles is crucial. The welfare of crew members, including their health, safety, and autonomy, must be safeguarded. Ethical frameworks should guide the decision-making process to ensure that interstellar exploration is conducted responsibly.

35. **Question**: How do current space missions contribute to the goal of interstellar travel?

    Answer: Current space missions contribute to the goal of interstellar travel by advancing technologies, scientific knowledge, and operational experience. Missions like NASA's Voyager and New Horizons have explored the outer solar system, providing valuable data on the interstellar medium. Research on propulsion systems, life support, and autonomous operations in missions like the Mars rovers and the ISS informs the development of interstellar capabilities. These missions lay the groundwork for future interstellar endeavors.

36. **Question**: What role does international collaboration play in interstellar exploration?

    Answer: International collaboration plays a crucial role in interstellar exploration by pooling resources, expertise, and technologies from multiple countries. Collaborative efforts can reduce costs, share risks, and accelerate the development of advanced propulsion and life support systems. International partnerships foster innovation and ensure that the benefits of interstellar exploration are shared globally. Cooperation among spacefaring nations is essential for achieving the ambitious goals of interstellar travel.

37. **Question**: What are the potential benefits of interstellar travel for humanity?

    Answer: The potential benefits of interstellar travel for humanity include the expansion of scientific knowledge, the discovery of new worlds, and the potential for long-term survival. Exploring other star systems can reveal insights into the formation and evolution of planetary systems and the potential for life beyond Earth. Interstellar travel also offers the possibility of establishing human settlements beyond the solar system, ensuring the survival of humanity in the face of existential threats.

38. **Question**: How does the concept of generation ships address the challenges of interstellar travel?

    Answer: The concept of generation ships addresses the challenges of interstellar travel by proposing self-sustaining spacecraft that support multiple generations of humans during the journey to distant stars. These ships would be equipped with advanced life support, agricultural, and recycling systems to sustain the crew over long durations. Generation ships offer a feasible solution for interstellar travel within the constraints of current propulsion technologies, allowing humanity to reach other star systems without the need for near-light-speed travel.

39. **Question**: What technological advancements are needed to make interstellar travel feasible?

    Answer: Technological advancements needed to make interstellar travel feasible include breakthroughs in propulsion, energy generation, life support, and autonomous systems. Developing propulsion systems capable of achieving high speeds, such as nuclear or antimatter propulsion, is critical. Energy generation technologies, such as fusion reactors, can provide the necessary power for long-duration missions. Advanced life support systems and AI-driven autonomous operations are essential for maintaining crew health and ensuring mission success.

40. **Question**: How does the search for exoplanets influence interstellar exploration efforts?

    Answer: The search for exoplanets influences interstellar exploration efforts by identifying potential targets for future missions. Discoveries of Earth-like exoplanets in habitable zones increase the likelihood of finding environments that can support life. Detailed studies of exoplanet atmospheres and surfaces provide valuable data for mission planning. The search for exoplanets drives technological advancements in telescopes and observational techniques, supporting the broader goals of interstellar exploration. (See Figure 7.4.)

FIGURE 7.4 The first direct image of an exoplanet, 2M1207b, orbiting the brown dwarf 2M1207. This image was taken by the Very Large Telescope (VLT) in 2004. Source: Wikimedia Commons.

41. **Question**: What is the significance of the discovery of water and organic molecules on other celestial bodies?
Answer: The discovery of water and organic molecules on other celestial bodies is significant because it indicates the potential for life and the presence of essential resources for human exploration. Water is crucial for life support, as it can be used for drinking, growing food, and producing oxygen and fuel. Organic molecules are the building blocks of life, suggesting that conditions conducive to life may exist beyond Earth. These discoveries inform the search for life and the planning of future missions.

42. **Question**: How can space telescopes contribute to interstellar exploration?
Answer: Space telescopes contribute to interstellar exploration by providing detailed observations of distant star systems and exoplanets. Telescopes like the Hubble Space Telescope, James Webb Space Telescope, and future missions can study the atmospheres, climates, and potential habitability of exoplanets. High-resolution imaging and spectroscopy reveal the composition and dynamics of celestial bodies. Space telescopes enhance our understanding of the universe and guide the selection of targets for interstellar missions.

43. **Question**: What are the potential impacts of discovering extraterrestrial life on interstellar exploration?
Answer: Discovering extraterrestrial life would have profound impacts on interstellar exploration by fundamentally changing our understanding of life in the universe and motivating further exploration. Such a discovery would raise new scientific questions and ethical considerations regarding the interaction with and protection of alien ecosystems. It would likely increase public interest and investment in space missions, driving advancements in technology and expanding the scope of exploration efforts.

44. **Question**: How do space missions to the outer solar system prepare us for interstellar travel?
Answer: Space missions to the outer solar system, such as Voyager, New Horizons, and future probes, prepare us for interstellar travel by testing technologies, conducting scientific research, and gathering data on the interstellar medium. These missions demonstrate the feasibility of long-duration space travel, autonomous operations, and deep space communication. They provide valuable insights into the challenges and opportunities of exploring beyond the solar system, laying the groundwork for future interstellar missions.

45. **Question**: What are the potential challenges of communicating with interstellar probes?

    Answer: Communicating with interstellar probes poses significant challenges due to the vast distances involved, leading to long communication delays and signal attenuation. Ensuring reliable data transmission requires advanced communication technologies, such as high-gain antennas, laser communication, and autonomous data processing. Developing robust protocols for delayed and intermittent communication is essential to manage the exchange of information between interstellar probes and Earth.

46. **Question**: How can international space agencies collaborate on interstellar missions?

    Answer: International space agencies can collaborate on interstellar missions by sharing expertise, resources, and technologies. Collaborative efforts can include joint development of propulsion systems, life support technologies, and scientific instruments. Shared funding and risk management enhance the feasibility and sustainability of interstellar missions. International agreements and partnerships, similar to those for the ISS, can facilitate coordination and ensure that the benefits of interstellar exploration are shared globally.

47. **Question**: What is the potential role of artificial intelligence in maintaining interstellar spacecraft?

    Answer: Artificial intelligence (AI) can play a crucial role in maintaining interstellar spacecraft by enabling autonomous monitoring, diagnostics, and repairs. AI systems can detect anomalies, optimize system performance, and execute maintenance tasks without human intervention. This capability is essential for long-duration missions where real-time communication with Earth is not possible. AI-driven maintenance ensures the reliability and longevity of interstellar spacecraft, increasing the chances of mission success.

48. **Question**: How can future generations be prepared for interstellar exploration?

    Answer: Future generations can be prepared for interstellar exploration through education, training, and public engagement. Educational programs that emphasize STEM fields and space sciences inspire interest and develop the skills needed for space missions. Training programs for astronauts and space engineers ensure that future crews are equipped to handle the challenges of interstellar travel. Public engagement initiatives, such as citizen science projects and space-related media, foster a culture of exploration and innovation.

49. **Question**: What is the significance of the discovery of interstellar objects passing through our solar system?
Answer: The discovery of interstellar objects passing through our solar system, such as 'Oumuamua and Comet Borisov, is significant because it provides direct evidence of material from other star systems. Studying these objects offers insights into the composition and dynamics of other planetary systems. These discoveries highlight the interconnectedness of star systems and the potential for interstellar exchange of materials. Analyzing interstellar objects informs our understanding of the broader universe and the potential for life beyond our solar system. (See Figure 7.5.)

FIGURE 7.5 The discovery of interstellar objects passing through our solar system, such as 'Oumuamua and Comet Borisov, is significant because it provides direct evidence of material from other star systems. Source: Wikimedia Commons.

50. **Question**: How does the concept of terraforming relate to interstellar exploration?
Answer: The concept of terraforming relates to interstellar exploration by envisioning the transformation of extraterrestrial environments to make them habitable for humans. Terraforming involves modifying the atmosphere, temperature, and surface conditions of a planet or moon to support human life. While primarily associated with Mars, terraforming concepts could be applied to exoplanets discovered during interstellar missions. Developing terraforming technologies presents significant scientific and ethical challenges but offers the potential for creating new habitable worlds.

51. **Question**: What are the potential long-term benefits of interstellar travel for humanity?

    Answer: The potential long-term benefits of interstellar travel for humanity include the expansion of human presence beyond the solar system, the discovery of new worlds and resources, and the advancement of scientific knowledge. Interstellar travel offers the possibility of establishing human settlements on distant planets, ensuring the survival and growth of human civilization. The pursuit of interstellar exploration drives technological innovation, fosters international collaboration, and inspires future generations to explore the cosmos.

52. **Question**: How can space missions to distant stars be funded and sustained?

    Answer: Funding and sustaining space missions to distant stars require a combination of government investment, private sector involvement, and international collaboration. Public funding from space agencies provides initial support for research and development. Private sector investment drives innovation and commercialization of space technologies. International partnerships share costs, resources, and expertise. Long-term sustainability depends on the successful demonstration of key technologies and the establishment of economic incentives for interstellar exploration.

53. **Question**: What are the ethical considerations of sending humans on interstellar missions?

    Answer: Ethical considerations of sending humans on interstellar missions include ensuring the safety, health, and well-being of the crew, addressing the potential impact on extraterrestrial environments, and considering the long-term implications for human society. The risks and benefits of interstellar missions must be carefully weighed, and ethical frameworks should guide decision-making. Protecting the rights and autonomy of crew members, as well as ensuring that exploration efforts are conducted responsibly and equitably, is essential.

54. **Question**: How will the quest for interstellar travel shape the future of space exploration?

    Answer: The quest for interstellar travel will shape the future of space exploration by driving advancements in propulsion, life support, and autonomous technologies. It will inspire new scientific discoveries, foster international collaboration, and push the boundaries of human knowledge and capability. The pursuit of interstellar travel represents humanity's aspiration to explore and understand the universe, ensuring that the spirit of exploration continues to thrive and evolve. As we strive to reach distant stars, the challenges and opportunities of interstellar travel will define the next era of space exploration.

The challenges and future prospects of space exploration encompass a wide range of technical, ethical, and societal considerations. Managing space debris and radiation, ensuring sustainable exploration and habitation, and pursuing the dream of interstellar travel are all critical aspects that will shape the trajectory of human civilization in the cosmos. Through innovation, collaboration, and a commitment to responsible exploration, humanity can overcome these challenges and achieve remarkable milestones in our journey through space. The advancements in technology, international partnerships, and ethical frameworks developed today will pave the way for a sustainable and inclusive future in space exploration. The pursuit of interstellar travel, in particular, represents humanity's aspiration to explore and understand the universe, ensuring that the spirit of exploration continues to thrive and evolve. As we strive to reach distant stars, the challenges and opportunities of interstellar travel will define the next era of space exploration, pushing the boundaries of human knowledge and capability.

REFERENCES

[Dick06] Dick, Steven J., and Roger D. Launius, eds., *Critical Issues in the History of Spaceflight*, NASA, 2006.

[Grinspoon16] Grinspoon, David, *Earth in Human Hands: Shaping Our Planet's Future*, Grand Central Publishing, 2016.

[Handberg03] Handberg, Roger, *Reinventing NASA: Human Space Flight, Bureaucratic Reform, and the Future of the U.S. Space Program*, Praeger, 2003.

[Impey15] Impey, Chris, and William F. Henry, *Beyond: Our Future in Space*, W.W. Norton & Company, 2015.

[Johnson-Freese07] Johnson-Freese, Joan, *Space as a Strategic Asset*, Columbia University Press, 2007.

[Logsdon96] Logsdon, John M., ed., *Exploring the Unknown: Selected Documents in the History of the U.S. Civil Space Program, Volume II: External Relationships*, NASA SP-4407, 1996.

[McCurdy97] McCurdy, Howard E., *Space and the American Imagination*, Smithsonian Institution Press, 1997.

CONCLUSION: REFLECTIONS ON THE JOURNEY AND THE ROAD AHEAD

REFLECTIONS ON THE JOURNEY

The journey through the exploration of space has been marked by remarkable achievements, profound discoveries, and unprecedented collaboration among nations. This book has taken us from the earliest human fascination with the cosmos to the cutting-edge technologies that promise to carry us to distant stars. Along the way, we have encountered the key milestones, scientific breakthroughs, and the visionary efforts of countless individuals who have made space exploration possible.

Chapter 1 "The Human Fascination with the Cosmos" explored the roots of our celestial curiosity, tracing the evolution of astronomy from ancient civilizations to the scientific revolution. Early astronomers laid the groundwork for our understanding of the universe, inspiring generations to look up and wonder about the mysteries of the stars. The myths and legends of the ancient Greeks, Egyptians, and other cultures show that our interest in the cosmos is deeply embedded in our history. The development of early astronomical tools and the formulation of basic celestial models set the stage for more advanced scientific inquiries.

Chapter 2 "The Dawn of the Space Age" delved into the intense competition and collaboration between the USA and the USSR during the Space Race. This era was defined by historic milestones such as the launch of Sputnik, the first human in space, Yuri Gagarin, and the monumental Apollo 11 moon landing. These achievements demonstrated humanity's potential to reach beyond our earthly confines and set the stage for future exploration. The Space Race was not just a demonstration of technological prowess but also a testament to human ingenuity and the desire to push boundaries. The triumphs and tragedies of this era, from the success of Apollo 11 to the losses of the Apollo 1 and Soyuz 1 crews, underscore the risks and rewards of space exploration.

Chapter 3 "Robotic Explorers Beyond Earth" highlighted the pivotal role of robotic missions in expanding our knowledge of the solar system. From Mars rovers like Curiosity and Perseverance to the Voyager and New Horizons missions, these robotic explorers have provided invaluable data about distant worlds and set the precedent for human missions. The Mars rovers have uncovered signs of past water activity, while Voyager's grand tour of the outer planets has revealed the diverse and dynamic nature of our solar system's giants. These missions have not only expanded our understanding of planetary science but also demonstrated the capabilities of modern robotics and artificial intelligence in conducting complex scientific investigations from millions of miles away.

Chapter 4 "Spacecraft and Technologies" examined the technological advancements that have made space travel possible. The evolution of rocket technology, spacecraft engineering from Mercury to Artemis, and innovations in propulsion and materials have all contributed to the growing capabilities of space exploration. The development of reusable rockets by companies like SpaceX has revolutionized the economics of space travel, making it more affordable and frequent. Advances in spacecraft materials and engineering have led to the creation of more durable and capable vessels, capable of withstanding the harsh conditions of space and ensuring the safety and comfort of astronauts on long-duration missions.

Chapter 5 "The International Space Station and Beyond" celebrated the ISS as a symbol of international cooperation and scientific progress. The ISS has been a hub for research, technology testing, and international partnerships, providing a model for future collaborative efforts in space exploration. The contributions of NASA, Roscosmos, ESA, JAXA, and CSA to the ISS demonstrate the power of global cooperation. The scientific experiments conducted on the ISS have led to breakthroughs in fields such as medicine, materials science, and environmental science. The ISS has also served as a crucial testbed for technologies that will be used in future deep space missions, including life support systems, radiation protection, and robotic systems.

Chapter 6 "Space for All: Commercial Ventures" discussed the transformative impact of private space companies on the industry. The rise of companies like SpaceX, Blue Origin, and Virgin Galactic has democratized access to space, introduced space tourism, and opened new possibilities such as asteroid mining. These companies are pushing the boundaries of what is possible in space travel, reducing costs through innovative technologies and business models. The development of space tourism is bringing the dream of space travel closer to reality for more people, while the prospect of asteroid mining offers new economic opportunities and the potential to source critical materials from space.

Chapter 7 "Challenges and Future Prospects" addressed the critical challenges facing space exploration, including space debris, radiation, and ethical considerations. It also explored sustainable exploration and habitation, as well as the ambitious quest for interstellar travel. These challenges require innovative solutions and international collaboration to ensure a responsible and prosperous future in space. The issue of space debris, for example, necessitates the development of effective debris mitigation and removal strategies to protect both current and future space missions. Radiation protection remains a significant hurdle for long-duration missions beyond Earth's protective magnetosphere, requiring advancements in shielding technologies and medical countermeasures.

THE ROAD AHEAD

As we look to the future, the road ahead for space exploration is filled with both challenges and opportunities. The continued advancement of technology, international cooperation, and the growing involvement of the private sector will drive progress in this field.

Sustainable exploration and habitation are paramount for the future of space exploration. Developing efficient resource utilization, closed-loop life support systems, and renewable energy sources is essential. Building sustainable habitats on the moon, Mars, and beyond will be crucial for long-term human presence in space. These habitats will need to be self-sufficient, relying on in-situ resource utilization (ISRU) technologies to extract water, oxygen, and building materials from the local environment. The integration of renewable energy sources such as solar and nuclear power will ensure a reliable and sustainable energy supply for these outposts.

International collaboration remains critical. The success of the ISS has shown that large-scale space projects benefit immensely from shared expertise, resources, and funding. Future missions to the moon, Mars, and beyond will thrive on collaborative efforts guided by frameworks like the Artemis Accords, which emphasize ethical and responsible exploration. Joint missions, such as the planned Lunar Gateway, will serve as staging points for lunar surface operations and provide a platform for scientific research and technology testing in deep space. Such international partnerships will also help to distribute the costs and risks associated with these ambitious endeavors, making them more feasible and sustainable.

Private space companies will continue to revolutionize space exploration. Companies like SpaceX, Blue Origin, and others are reducing launch costs, increasing access to space, and developing new technologies. The commercial space sector will expand opportunities for scientific research, tourism, and resource extraction, making space more accessible. The development of reusable launch vehicles and innovative spacecraft designs are driving down the cost of access to space, making it possible for a wider range of actors, from national space agencies to private companies and academic institutions, to participate in space missions. The growth of space tourism will provide new revenue streams and inspire public interest and support for space exploration.

Addressing space debris and environmental concerns is vital for protecting the space environment. Technological solutions for debris removal, sustainable launch practices, and international regulations are necessary to ensure safe and sustainable space operations. The development of active debris removal (ADR) technologies, such as robotic arms, nets, and laser systems, will help to mitigate the threat posed by space debris. Implementing stricter guidelines for satellite deorbiting and end-of-life disposal will also be crucial in preventing the accumulation of new debris. Sustainable launch practices, including the use of greener propulsion technologies and minimizing the environmental impact of rocket launches, will be essential for preserving the space environment for future generations.

Human missions to Mars and beyond are becoming increasingly feasible. Advancements in propulsion, life support, and habitat technologies are paving the way for these ambitious endeavors. Careful planning, robust technologies, and international cooperation will be essential for the success of these missions. Developing propulsion technologies, such as nuclear thermal propulsion and ion thrusters, will enable faster and more efficient travel to Mars and other distant destinations. Advanced life support systems, capable of recycling air, water, and waste, will be critical for sustaining human life on long-duration missions. Habitats designed to protect against radiation and provide a comfortable living environment will ensure the health and well-being of astronauts.

Ethical considerations and planetary protection must guide our exploration efforts. Ensuring that our activities do not harm extraterrestrial environments or jeopardize Earth's biosphere is crucial. International agreements and ethical frameworks will ensure responsible exploration. The principles of planetary protection, as outlined by the Committee on Space Research (COSPAR), provide guidelines for preventing biological contamination of other celestial bodies and safeguarding Earth's biosphere from potential extraterrestrial organisms. These principles must be adhered to in all space missions to ensure the integrity of scientific research and the safety of our planet.

Space exploration has the power to inspire and educate. Engaging the public, especially the younger generation, through education and outreach programs will foster curiosity and innovation. Encouraging diverse participation in space sciences will ensure that the benefits of space exploration are shared globally. Educational initiatives, such as STEM programs and citizen science projects, will inspire the next generation of scientists, engineers, and explorers. Public outreach efforts, including media coverage, public talks, and interactive exhibits, will help to raise awareness and support for space exploration.

In conclusion, the journey through space exploration is far from over. The achievements of the past and present serve as a foundation for the ambitious goals of the future. By embracing innovation, fostering international collaboration, and addressing the challenges ahead, humanity can continue to explore, discover, and thrive in the vast expanse of space. The road ahead is filled with promise, and the pursuit of knowledge and exploration will remain a defining aspect of our shared human experience. As we venture further into the cosmos, we will continue to push the boundaries of what is possible, uncovering new mysteries and expanding our understanding of the universe. The journey to the stars is a testament to the indomitable spirit of human curiosity and our enduring desire to explore the unknown.

GLOSSARY

Apollo program: NASA's program of space missions that aimed to land humans on the Moon and bring them safely back to Earth. Significant missions include Apollo 11, which achieved the first Moon landing in 1969. This program set the stage for all subsequent human space exploration efforts by demonstrating the feasibility of sending humans to another celestial body and returning them safely. The Apollo program also included significant scientific experiments and lunar sample collection that advanced our understanding of the Moon's composition and history. (See Chapter 2 "The Dawn of the Space Age.")

Artemis program: NASA's current lunar exploration program aiming to return humans to the Moon and establish a sustainable presence by the end of the decade. The program includes the development of the Space Launch System (SLS) and the Orion spacecraft. Artemis also aims to foster international partnerships and pave the way for future human missions to Mars. The program focuses on inclusivity, aiming to send the first woman and the next man to the lunar surface. (See Chapter 4 "Spacecraft and Technologies" and Chapter 5 "The International Space Station and Beyond.")

asteroid mining: The process of extracting valuable minerals and other resources from asteroids. Companies like Planetary Resources and Deep Space Industries are pioneering efforts in this field. The potential for asteroid mining to provide resources for both space missions and terrestrial use makes it a significant focus of future commercial space ventures. These resources include precious metals, water for life support and fuel, and other materials critical for in-situ manufacturing in space. (See Chapter 6 "Space for All: Commercial Ventures.")

astronomical unit (AU): A unit of distance used in astronomy, equivalent to the average distance between the Earth and the Sun, approximately 93 million miles or 150 million kilometers. It is used to express distances within our solar system.

astrobiology: The study of the origin, evolution, distribution, and future of life in the universe. This interdisciplinary field combines elements of biology, chemistry, and geology to understand the potential for life on other planets and moons. Astrobiology research is integral to missions like those of the Mars rovers, which search for signs of past life on Mars. (See Chapter 3 "Robotic Explorers Beyond Earth.")

cosmic rays: High-energy particles from outer space that can pose a radiation hazard to astronauts and spacecraft. The impact of cosmic rays on human health and spacecraft electronics is a major concern for long-duration space missions, such as those to Mars or beyond. Understanding and mitigating cosmic ray exposure is crucial for the safety of astronauts on deep space missions. (See Chapter 7 "Challenges and Future Prospects.")

CubeSat: A type of miniaturized satellite used for space research, typically measuring 10x10x10 centimeters. CubeSats have revolutionized space exploration by making it more accessible and affordable for educational institutions, small businesses, and developing countries. They are used for a variety of purposes, including Earth observation, scientific experiments, and technology demonstrations. (See Chapter 6 "Space for All: Commercial Ventures.")

exoplanet: A planet located outside our solar system. The search for exoplanets has been a major driver of advances in telescopes and observational techniques, influencing interstellar exploration efforts and the quest to find habitable worlds beyond our own. The discovery of exoplanets in the habitable zone of their stars has profound implications for the potential existence of extraterrestrial life. (See Chapter 7 "Challenges and Future Prospects.")

Falcon 9: A reusable rocket developed by SpaceX. It has revolutionized the space industry by reducing launch costs and increasing the frequency of missions. The Falcon 9's ability to land and be reused has set a new standard for commercial spaceflight. It is used for a variety of missions, including satellite launches, cargo resupply missions to the ISS, and crewed spaceflights. (See Chapter 6 "Space for All: Commercial Ventures.")

geostationary orbit: A circular orbit around Earth in which a satellite 's orbital period matches the Earth's rotation period. This allows the satellite to remain fixed over one point on the Earth's surface, making it ideal for communication and weather satellites.

Hohmann transfer orbit: An efficient way of moving a spacecraft between two orbits using two engine burns. It is commonly used for missions traveling between Earth and other planets or between different orbits around the same planet.

International Space Station (ISS): A space station, or habitable artificial satellite, in low Earth orbit. It serves as a hub for scientific research and international collaboration. The ISS is a testament to what can be achieved through global cooperation, providing invaluable insights into living and working in space. It is a joint project involving space agencies from the United States, Russia, Europe, Japan, and Canada. (See Chapter 5 "The International Space Station and Beyond.")

in-situ resource utilization (ISRU): The practice of collecting and using materials found or manufactured on other astronomical objects (such as the Moon or Mars) to support space missions. ISRU is critical for reducing the cost and complexity of space missions by utilizing local resources instead of transporting everything from Earth. This includes extracting water, oxygen, and building materials from lunar or Martian soil. (See Chapter 7 "Challenges and Future Prospects.")

Kessler Syndrome: A scenario in which the density of objects in low Earth orbit is high enough to cause collisions that generate more space debris, increasing the likelihood of further collisions. This presents a significant challenge for space operations and highlights the need for effective space debris management strategies. The accumulation of debris can create a cascading effect that threatens the safety and sustainability of space activities. (See Chapter 7 "Challenges and Future Prospects.")

Mars rover: Robotic vehicles designed to explore the surface of Mars. Notable examples include Curiosity and Perseverance, which have provided invaluable data about the Red Planet's geology, climate, and potential for past life. These rovers are equipped with scientific instruments to conduct experiments, analyze soil and rock samples, and search for signs of water and life. (See Chapter 3 "Robotic Explorers Beyond Earth.")

nuclear thermal propulsion: A proposed method of spacecraft propulsion that uses a nuclear reactor to heat propellant to generate thrust. This technology could significantly reduce travel time for interstellar missions, making distant space exploration more feasible. Nuclear thermal rockets offer higher efficiency than chemical rockets and are considered for missions to Mars and beyond. (See Chapter 7 "Challenges and Future Prospects.")

orbital mechanics: The study of the motions of spacecraft and celestial bodies under the influence of gravitational forces. It involves calculating trajectories, orbits, and maneuvers needed for missions to reach their destinations.

planetary Protection: The practice of protecting solar system bodies from contamination by Earth life and protecting Earth from possible life forms that may be returned from other solar system bodies. Planetary protection is critical for ensuring the integrity of scientific research and the safety of Earth's biosphere. It involves strict protocols for sterilizing spacecraft and handling samples returned from other planets. (See Chapter 7 "Challenges and Future Prospects.")

Proxima Centauri: The closest known star to the Sun, located about 4.24 light-years away. It is part of the Alpha Centauri star system and has been a target of interest for potential interstellar missions, such as the Breakthrough Starshot initiative. (See Chapter 7 "Challenges and Future Prospects.")

reusable rocket technology: Technology that allows rockets to be used multiple times, significantly reducing the cost of space launches. SpaceX's Falcon 9 and Blue Origin's New Shepard are notable examples that have set new benchmarks in the industry. Reusable rockets are crucial for making space travel more economical and sustainable. (See Chapter 6 "Space for All: Commercial Ventures.")

solar sail: A method of propulsion that uses large, reflective sails to capture the momentum of sunlight for propulsion. Solar sails are being developed for missions to explore the outer solar system and beyond, offering a fuel-free method of continuous acceleration. The concept relies on the pressure exerted by photons from the Sun to propel the spacecraft. (See Chapter 7 "Challenges and Future Prospects.")

space debris: Nonfunctional, human-made objects in space, such as defunct satellites and spent rocket stages. Space debris poses a significant threat to operational satellites and spacecraft, necessitating the development of mitigation and removal strategies. Effective space debris management is essential for the safety and sustainability of space operations. (See Chapter 7 "Challenges and Future Prospects.")

space tourism: Commercial activities that allow private individuals to travel to space for leisure. Companies like Virgin Galactic and Blue Origin are pioneers in this field, offering suborbital flights that provide passengers with a few minutes of weightlessness and spectacular views of Earth. Space tourism is expected to grow, with companies planning orbital and lunar missions for paying customers. (See Chapter 6 "Space for All: Commercial Ventures.")

SpaceX: A private aerospace manufacturer and space transportation company founded by Elon Musk. SpaceX has revolutionized the space industry with innovations such as the Falcon 9 reusable rocket and the Crew Dragon spacecraft, significantly lowering the cost of access to space. The company's achievements include delivering cargo to the ISS, launching numerous satellites, and planning crewed missions to Mars. (See Chapter 6 "Space for All: Commercial Ventures.")

Van Allen radiation belts: Zones of charged particles trapped by Earth's magnetic field. These belts pose a radiation hazard to spacecraft and astronauts, especially during missions beyond low Earth orbit. Understanding and mitigating the effects of radiation from the Van Allen belts is crucial for the safety of human spaceflight.

Voyager missions: NASA's space missions launched in 1977 to explore the outer planets and beyond. The Voyager probes have provided invaluable data on the gas giants and are now traveling through interstellar space, continuing to send back data as they move farther from the Sun. The Voyager missions have expanded our understanding of the outer solar system and the interstellar medium. (See Chapter 3 "Robotic Explorers Beyond Earth.")

TIMELINE OF KEY EVENTS

1957: Launch of Sputnik 1 by the USSR, marking the beginning of the Space Age. This historic event initiated the space race between the United States and the Soviet Union and led to significant advancements in space technology and exploration. The launch of Sputnik was not just a technical milestone; it also had profound political and social implications, sparking widespread public interest and leading to the establishment of NASA. (See Chapter 2 "The Dawn of the Space Age.")

1958: The United States launches its first satellite, Explorer 1, marking its entry into the Space Race. This mission also led to the discovery of the Van Allen radiation belts, a major scientific achievement.

1961: Yuri Gagarin becomes the first human in space aboard Vostok 1. His successful orbit of Earth demonstrated the feasibility of human spaceflight and paved the way for future manned missions. Gagarin's mission was a pivotal moment in the Space Race, showcasing Soviet space prowess and intensifying American efforts to catch up. (See Chapter 2 "The Dawn of the Space Age.")

1962: John Glenn orbits Earth in Friendship 7, becoming the first American to do so. This mission was a significant achievement for NASA and helped restore American confidence in its space program.

1963: Valentina Tereshkova becomes the first woman in space, flying aboard Vostok 6. Her mission was a significant milestone in the history of human spaceflight, highlighting the Soviet Union's commitment to gender equality in space exploration.

1965: Alexei Leonov performs the first spacewalk during the Voskhod 2 mission, demonstrating new capabilities in human spaceflight. This was another great success for the Soviet Union and highlighted the complexities and dangers of extravehicular activities (EVAs).

1966: The Soviet Luna 9 mission achieves the first soft landing on the Moon, providing valuable data about the lunar surface. This mission was a precursor to later manned lunar landings and underscored the technological advancements of the Soviet space program.

1969: Apollo 11 mission lands the first humans on the Moon. Neil Armstrong and Buzz Aldrin's historic moonwalk fulfilled President Kennedy's goal of landing a man on the Moon and returning him safely to Earth, inspiring generations of scientists and explorers. This mission represented the pinnacle of the Space Race and a triumph of human ingenuity and perseverance. (See Chapter 2 "The Dawn of the Space Age.")

1971: Launch of Salyut 1, the world's first space station, by the Soviet Union. This marked the beginning of long-duration human habitation in space and set the stage for future space stations like Mir and the ISS.

1973: Launch of Skylab, the United States' first space station, which conducted important scientific research and technological demonstrations. Skylab provided valuable experience in living and working in space for extended periods.

1977: Launch of the Voyager missions to explore the outer planets and beyond. These missions have provided detailed data on Jupiter, Saturn, Uranus, and Neptune, as well as their moons and rings, and are now traveling through interstellar space (see Chapter 3: Robotic Explorers Beyond Earth).

1981: The first Space Shuttle flight (STS-1) inaugurates a new era of reusable spacecraft. The Space Shuttle program significantly advanced our ability to conduct scientific research and deploy satellites, and it played a key role in building the International Space Station. The Shuttle's reusability was a revolutionary concept that reduced the cost of access to space and enabled a wide range of missions. (See Chapter 4 "Spacecraft and Technologies.")

1986: The Challenger disaster occurs, leading to a reevaluation of safety protocols in space missions. This tragic event underscored the inherent risks of space travel and prompted significant changes in NASA's approach to crew safety.

1990: Launch of the Hubble Space Telescope, which has provided unprecedented views of the universe and advanced our understanding of astronomy. Hubble's discoveries have transformed our knowledge of the cosmos, from the expansion rate of the universe to the formation of galaxies.

1998: Launch of the first module of the International Space Station (ISS), beginning its construction. The ISS represents one of the most significant achievements in international collaboration and continues to serve as a platform for scientific research and technology development. The station's construction and continuous operation have provided invaluable insights into long-duration human spaceflight and international teamwork. (See Chapter 5 "The International Space Station and Beyond.")

2004: SpaceShipOne, funded by the Ansari X Prize, becomes the first privately funded spacecraft to reach space. This milestone demonstrated the viability of private spaceflight and set the stage for the development of commercial space tourism. SpaceShipOne's success showed that nongovernmental entities could participate in and contribute to space exploration. (See Chapter 6 "Space for All: Commercial Ventures.")

2008: SpaceX's Falcon 1 becomes the first privately developed liquid-fueled rocket to reach orbit, marking a significant achievement for commercial spaceflight. This milestone was a precursor to the more advanced Falcon 9 and laid the groundwork for SpaceX's future successes.

2010: SpaceX's Dragon spacecraft completes its first successful test flight. This event marked the beginning of a new era in commercial space transportation, with SpaceX becoming the first private company to send a spacecraft to the International Space Station. Dragon's success demonstrated the potential of commercial partnerships in achieving space exploration goals. (See Chapter 6 "Space for All: Commercial Ventures.")

2011: The end of the Space Shuttle program. The retirement of the Space Shuttle marked the end of an era in human spaceflight and led to the development of new spacecraft and commercial partnerships for crewed missions. This transition highlighted the need for new approaches to space transportation and the importance of international collaboration. (See Chapter 4 "Spacecraft and Technologies.")

2012: Curiosity rover lands on Mars, beginning a mission to explore the planet's surface. Curiosity's discoveries have provided significant insights into Mars' geology, climate, and potential for past life, laying the groundwork for future human missions. The rover's advanced suite of scientific instruments has enabled detailed studies of the Martian environment. (See Chapter 3 "Robotic Explorers Beyond Earth.")

2015: New Horizons mission conducts a flyby of Pluto, providing the first close-up images of the dwarf planet. This mission expanded our understanding of the outer reaches of our solar system and the diverse characteristics of its objects. New Horizons' findings have reshaped our knowledge of Pluto and its moons. (See Chapter 3 "Robotic Explorers Beyond Earth.")

2016: Launch of the Breakthrough Starshot initiative to develop technologies for interstellar travel. This ambitious project aims to send small, lightweight spacecraft to the nearest star system, Alpha Centauri, within a generation. Breakthrough Starshot represents a bold vision for the future of space exploration and the quest to reach other star systems. (See Chapter 7 "Challenges and Future Prospects.")

2020: Launch of the Perseverance rover to Mars, continuing the exploration of the Red Planet. Perseverance is tasked with searching for signs of past microbial life and collecting samples for future return to Earth. The rover's advanced technology and scientific objectives mark a new phase in Mars exploration. (See Chapter 3 "Robotic Explorers Beyond Earth.")

2021: SpaceX's Inspiration4 mission becomes the first all-civilian crewed spaceflight. This mission demonstrated the potential for ordinary citizens to participate in space travel and highlighted the growing role of private companies in human spaceflight. Inspiration4's success showcased the democratization of space access and the potential for future commercial crewed missions. (See Chapter 6 "Space for All: Commercial Ventures.")

2023: Artemis I mission successfully completes its crewless flight around the Moon, marking a significant milestone in NASA's Artemis program. This mission is a critical step towards returning humans to the lunar surface and establishing a sustainable presence on the Moon. Artemis I's success lays the groundwork for future crewed Artemis missions and the long-term exploration of the Moon and beyond. (See Chapter 4 "Spacecraft and Technologies" and Chapter 7 "Challenges and Future Prospects.")

Additional Resources

BOOKS

- *The Right Stuff* by Tom Wolfe: A classic account of the early days of the American space program, focusing on the astronauts of the Mercury program. This book provides an in-depth look at the courage and determination of the first American astronauts. (See Chapter 2 "The Dawn of the Space Age.")

- *The Case for Mars* by Robert Zubrin: An influential book advocating for the human exploration and settlement of Mars, discussing the technological and economic feasibility of such missions. Zubrin's vision has inspired many in the space community to consider Mars as the next frontier for human exploration. (See Chapter 7 "Challenges and Future Prospects.")

- *Riding Rockets* by Mike Mullane: A memoir by a Space Shuttle astronaut that provides an inside look at the life of an astronaut and the operations of the Space Shuttle program. Mullane's candid and often humorous account offers a unique perspective on the triumphs and challenges of human spaceflight. (See Chapter 4 "Spacecraft and Technologies.")

WEB SITES

- International Space Station (*www.nasa.gov/mission_pages/station/main/index.html*): A comprehensive resource for information on the ISS, including current experiments, crew activities, and international partnerships. The ISS website offers detailed information on the station's construction, scientific contributions, and the experiences of astronauts living and working in space. (See Chapter 5 "The International Space Station and Beyond.)

- NASA (*www.nasa.gov*): The official Web site of the National Aeronautics and Space Administration, providing up-to-date information on current missions, research, and educational resources. NASA's extensive archive of images, videos, and documents offers a wealth of information on the history and future of space exploration (referenced throughout the book).

- SpaceX (*www.spacex.com*): The official website of SpaceX, offering information on their missions, technology, and future plans. SpaceX's innovative approach to space travel and their detailed mission profiles provide valuable insights into the commercial space industry. (See Chapter 6 "Space for All: Commercial Ventures.")

DOCUMENTARIES AND FILMS

- *Apollo 11* (2019, Universal Pictures): A documentary film that provides a detailed look at the Apollo 11 mission using archival footage and audio recordings. This film offers an immersive experience of the historic mission that landed the first humans on the Moon. (See Chapter 2 "The Dawn of the Space Age.")

- *The Mars Generation* (2017, Netflix): A documentary exploring the future of space exploration through the eyes of teenagers attending Space Camp. This film highlights the aspirations of the next generation of space explorers and the importance of education in STEM fields. (See Chapter 3 "Robotic Explorers Beyond Earth" and Chapter 7 "Challenges and Future Prospects.")

- *For All Mankind* (1989, National Geographic): A documentary that chronicles the Apollo missions to the Moon, featuring footage from NASA's archives and interviews with astronauts. This film captures the excitement and challenges of the Apollo era and the human spirit of exploration. (See Chapter 2 "The Dawn of the Space Age.")

ONLINE COURSES

- "Introduction to Aerospace Engineering: Astronautics and Human Spaceflight" (MIT OpenCourseWare): A free online course that covers the fundamentals of astronautics and human spaceflight. This course provides a comprehensive overview of the engineering principles and challenges involved in designing spacecraft and conducting space missions (See Chapter 4 "Spacecraft and Technologies.")

- "Space Exploration: From the Moon to Mars and Beyond" (Coursera): An online course offered by the University of Arizona that explores the history and future of space exploration. This course covers key topics in space science and technology, including lunar and Mars missions (referenced throughout the book).

- "Astrobiology and the Search for Extraterrestrial Life" (Coursera): A course by the University of Edinburgh that examines the scientific search for life beyond Earth. This course provides insights into the methods and discoveries of astrobiology, a field that is crucial for understanding the potential for life on other planets. (See Chapter 7 "Challenges and Future Prospects.")

RESEARCH PAPERS AND ARTICLES

- "The Next 100 Years of Planetary Exploration" by David W. Beaty et al.: A comprehensive review of the future directions and goals of planetary exploration. This paper outlines the scientific and technological priorities for exploring our solar system in the coming century. (See Chapter 3 "Robotic Explorers Beyond Earth" and Chapter 7 "Challenges and Future Prospects.")

- "Technological and Scientific Aspects of Solar Sails" by Giovanni Vulpetti et al.: An in-depth analysis of the potential and challenges of solar sail technology for space exploration. This paper discusses the design, propulsion mechanics, and mission applications of solar sails. (See Chapter 7 "Challenges and Future Prospects.")

- "The Artemis Accords: Advancing Human Exploration in the 21st Century" by NASA: A detailed explanation of the principles and guidelines outlined in the Artemis Accords for international cooperation in lunar exploration. This document highlights the importance of collaboration and ethical conduct in space missions. (See Chapter 5 "The International Space Station and Beyond" and Chapter 7 "Challenges and Future Prospects.")

INDEX

www.ingramcontent.com/pod-product-compliance
Lightning Source LLC
Chambersburg PA
CBHW061318220326
41599CB00026B/4944